A PLUME BOOK

A CALM BRAIN

GAYATRI DEVI, M.D., is a neurologist and the director of the New York Memory and Healthy Aging Services. A clinical associate professor at the NYU School of Medicine, she is the president of the not-for-profit National Council on Women's Health. She lives in New York City.

Praise for *A Calm Brain*

"A welcome alternative approach to overtaxing our brains and then reaching for the pill bottle—should warrant serious attention." —*Kirkus Reviews*

"*A Calm Brain* brings the power of cutting-edge neuroscience to everyday life. For anyone who wants to take charge of the twenty-first century while remaining calm, focused, and productive—this is the book for you."
—Henry S. Lodge, M.D., author of *The New York Times* bestseller *Younger Next Year*

"*A Calm Brain* shows readers why the brain craves calm, and how this will improve your health and happiness. Blending stories, science, and practical advice, it offers a path to a calmer life." —Paul J. Zak, author of *The Moral Molecule*

"Summer's supposed to be the time when you shift into vacation mode and slow down, but if you can't, this neurologist author offers insights into managing stress and much more."

—*Los Angeles Times*,
Summer Reading List Pick

A Calm Brain

How to Relax into a Stress-Free, High-Powered Life

Gayatri Devi, M.D.

A PLUME BOOK

PLUME
Published by the Penguin Group
Penguin Group (USA) Inc., 375 Hudson Street, New York, New York 10014, USA

USA | Canada | UK | Ireland | Australia | New Zealand | India | South Africa | China
Penguin Books Ltd, Registered Offices: 80 Strand, London WC2R 0RL, England
For more information about the Penguin Group visit penguin.com

First published in the United States of America by Dutton,
a member of Penguin Group (USA) Inc., 2012
First Plume Printing 2013

THE LIBRARY OF CONGRESS HAS CATALOGUED THE DUTTON EDITION AS FOLLOWS:

A calm brain : unlocking your natural relaxation system / Gayatri Devi.—1st ed.
p. cm.
Includes bibliographical references and index.
ISBN 978-0-525-95269-5 (hc.)
ISBN 978-0-14-219686-1 (pbk.)
1. Cognitive-analytic therapy. 2. Relaxation. I. Title.
RC489.C6D48 2012 613.7'92—dc23 2011046555

Printed in the United States of America
10 9 8 7 6 5 4 3 2 1

For Ginny and Mike

Acknowledgments

I would like to thank, in a *very* particular order:

Laura Yorke, literary agent extraordinaire: Expert horsewoman and inspiringly resilient recoverer from head injury.

Amy Hertz, Jo Ann Miller, and *Stephen Morrow*: My editing trio, who helped me give voice to my muddled thoughts.

Dani Levine and *Eamonn Vitt*: For their generous help with manuscript preparation.

William Blessing, Horacio Kaufmann, Joseph LeDoux, Randolph Nesse, and *Clifford Saper*: Neuroscientists from the universities of Flinders (Australia), New York, Michigan, and Harvard, for being kind and extraordinarily erudite sounding boards.

Roger Cracco, M.D.: Chairman Emeritus of Neurology, SUNY Downstate, who got me excited about the brain from the synapses on up.

Mahendra Somasundarum, M.D.: Distinguished Professor of Neurology, SUNY Downstate, who taught me that neurology is, indeed, the Queen of Medicine.

Lewis Glickman, M.D.: Professor Emeritus of Psychiatry, SUNY Downstate, who taught me to *listen* to patients.

Contents

A Calm Brain

INTRODUCTION

My Grandfather's Secret

Have you ever known someone with whom you feel incredibly at peace? Someone whose calm is a contagion that spreads and takes hold and soothes the most tempestuous of moods? A cop who is fantastic in a crisis and is the partner everyone wants? A go-to colleague who makes you feel better when your boss is throwing a tantrum? The friend you instinctively turn to if you are scared or jittery? I am not talking about someone who is boring or sedated, but rather someone who is in a state of focused, relaxed, and confident alertness, the optimal state in which to sail through the oceans of our lives.

I knew such a person. In the tempest of our family, my grandfather was the eye of the storm. All around him churned the chaotic entropy that is the Indian household. And at the center stood my Thatha, a monument of stalwart self-possession.

Thatha awoke at five every morning and made himself a cup of hot water. Then he sat down in an old slingback chair. I would perch on a stool next to him. We would both read the morning newspaper—or at least I would pretend to. When he was finished with the paper, he would stand up, cross his hands behind his back, and pace slowly up and down the length of the house in the tropical dawn. I would follow closely behind, not close enough to be a pest but close enough to observe and imitate his every movement. This would go on for about a half hour. He said it helped him think. There was a slow cadence to everything Thatha did during those morning hours before the household stirred and came alive.

My grandfather was a powerful man, a mathematician who went on to run the regional telephone company, in charge of thousands of employees. His quiet ways and demeanor belied the tremendous pressures and responsibilities that were part of his job. I remember him getting angry just once, when I was seven years old. As a joke, I dumped a bucket of manure into the well at the family farm, where we spent my school holidays. The farm had no drinking water as a result. My grandmother yelled and carried on, which fazed me not one bit. She was a volatile woman and prone to the excesses of temper characteristic of the many adults who swirled around my young life.

When my grandfather learned what had happened, the first thing he did was make arrangements to procure clean water. I stood there ashamed. I was his favorite grandchild.

He had never been angry with me before, but he looked furious now. When the clamor died down, the two of us were left alone in the garden. I looked up at him, terrified. He didn't say a word, just watched me gently as if evaluating what to say. Finally, he spoke. "That was an asinine thing to do," he said, so softly that I almost didn't hear him. And that was that. No further repercussions. I was devastated. To this day, few words have hurt me more.

Why? Because they came from this man I adored, this being of calm, this model of equanimity. I admired my grandfather in that moment, even in my mortification. That he attended to what needed to be attended to, that he reprimanded me succinctly and well, that he did all this so very calmly, and that, most of all, he communicated to his small, adoring granddaughter that everything was going to be okay.

At the time, I didn't quite grasp my grandfather's influence on me. I didn't yet know that I would try to spend much of my early adult life searching for the holy grail of his equanimity. In time I would learn the secrets of my grandfather's serenity. When, years later, I finally came upon the answer, I stumbled into it. Serendipitously. I discovered what my grandfather had, what I could have, that would allow me this delicious stillness, this confidence in times of tribulation, this calm. I went into medicine because it was my family trade. Both my great-grandfathers were physicians. My father was a surgeon, my mother a microbiologist. But within medicine, I chose my field—

neurology—because nothing fascinates me more than the brain and because I knew that the answer to my quest lay within it. In this book, I will share what I've learned about the neuroscience of calm and what we need to do to attain it. But first, let's look more closely at what I mean when I speak of "calm."

Calm is a sense of internal composure that lets us function to the best of our abilities. It is the ideal state of the brain, supported by a body completely allied with this purposeful brain, allowing us to harness our cognitive powers while maintaining a perfect balance with our emotions. When you are calm, you are in your zone, unperturbed by distractions or distress. How can we get there? What does it take to reach this state of optimal brain functioning?

The crucial turning point in my thinking about calm came when I realized that it is as much a neurological state and as much a product of the brain as are language or movement. What this means is that, under the best of circumstances, you can choose your emotional state just as you can choose to move your arm or tell a joke. And you can choose calm over panic in confronting a stressful situation. Why is this so? Because in the exquisite mechanism that is our body, every force has an equal and opposite force. The brain has complex systems for relaxation and calm to counteract its intricate mechanisms for alertness and anxiety. Therefore, we must look deep into the workings of the brain and the nervous system to uncover the keys to calm.

The answer to "How can we get there?" lies not within

our frontal lobes, our rational higher brain, the seat of logic and thinking, but within our core brain, which controls our emotions and impulses, and the vast environmental sensor and receptacle that is our body. My grandfather knew this. All his life, in his habits and his thoughts, he was deliberately augmenting the hardwired circuits between his body and his brain that promote—yes, promote—calm. The way he started each morning, the slow pouring out of his mind and body into the day, was worlds apart from the typical modern jolting awake and bolting into action. Throughout his life, Thatha was preferentially enhancing the connections between the calm centers in his body and in his brain. This served him well through hectic days—and it served him superbly when his little granddaughter did something idiotic.

The balance of calm and anxiety in my grandfather, and in all of us, is directed by the parasympathetic relaxing system and the sympathetic alerting system. Both of these carry information from the body into the core brain, which then turns up or turns down the calm "switch." Now consider what happens to these systems when we're tossed into the turmoil of an Internet-driven, multitasking, multi-roled, BlackBerryed life. Our primitive core brain interprets the buzzing and beeping, the flashing assaults on our attention, as if they were hungry saber-toothed tigers lurking on our path. Our brains, evolved to fight real threats like tigers, cannot distinguish between a ferocious beast poised to attack and a quarterly report that's two weeks late. Faced with a barrage of stimulation, guess which brain system we preferentially recruit. We invariably amp up our arousing sympathetic system, just as our ancestors drew on that system to help them escape or slay the tiger.

Our rational brain's arousing system may help us get that report finished and even handle ten other assignments at the same time, but the next task looms and then the next. And along the way, we forget how to turn down this ever-vigilant, worrying system. Fueled by the fears of the frontal lobes, which equate a missed deadline with the loss of a job and prestige, the sympathetic system creates an orange alert in our bodies and brains, a state of inchoate and perpetual anxiety. We wage a brain war without the intervening benefits of peace.

It is the charged conflict between the primitive core brain and the more evolved rational brain that leads to the angst and unrest that characterizes so much of modern life. We are unable to soothe ourselves. We can't sleep at night and we're tense and edgy during the day. We cannot concentrate, cannot relax, even during a massage or while lolling on a Caribbean beach. When we don't activate our body's calming parasympathetic mechanism, our runaway sympathetic system causes us to succumb to heart attacks and strokes, even as it helps us meet our daily deadlines.

But what if we made a different choice? A counterintuitive selection? What if instead of choosing the task-driven sympathetic system, we were to recruit the parasympathetic relaxing system? Paradoxically, as we will see, this choice may actually give us a better shot at winning the rat race, allowing us to make our deadlines *and* stay healthy. In fact, learning to make this choice may well keep us alive. At peace. And sleeping. Without Ambien.

It took me a while to understand this. All my life I have known that I function best in states of emergency. As a resident in training in a busy hospital, when there was a cardiac arrest or a gunshot wound, my brain would slow down and pick out the relevant information in crisp detail, allowing me clarity in milliseconds. Although I continued to think, I was almost on autopilot. Surprisingly, for someone prone to the dramatic—a trait I inherited from my grandmothers—I was at my calmest and most at peace during these emotionally heightened times.

In those days, I drank two six-packs of caffeinated Diet Coke and went through a pack of cigarettes each day. Every case was a race, every save a win. Some days, after such hectic nights, I would head out for a manicure. My hands would tremble from the excess of adrenaline in my system, and my manicurist would say, "Hmmm. . . . Guess who's just come off call!"

I was in sympathetic overdrive through much of my training, and it served me well. Because I was young, I could still ramp up my parasympathetic system and fall into a relaxing and deep sleep, trembles and all. Interestingly, though, if I was calmest in overdrive, I was more irritable during the less stressful times of my life. I would find myself thinking less clearly, torn between the demands of work and home. It was hard to take pleasure in the quiet of the dawn. I longed for the calm that suffused my grandfather and wanted this glow to settle down on me, like a blanket of power. I was stuck in this rut until Wolfie chanced into my life.

I found Wolfie one cold winter day as I was driving on the Long Island Expressway. He was wandering, shivering, on the median of the highway. He was a medium-size dog, part German shepherd and part collie, scrawny and scared. I pulled over to the side of the road and managed to get him into the car. When I brought him home, he quickly took to our family like Orphan Annie to Daddy Warbucks. There was just one problem. Every time he got into the car, Wolfie became excited, bouncing comically from window to window. By the end of every car trip, no matter how short, Wolfie, exhausted and anxious, would vomit all over the car.

We tried everything we could think of, but nothing seemed to work. At the end of every car ride, I was left cleaning the upholstery. Finally, I took Wolfie to see Ellen, a talented dog trainer. Ellen listened intently to my story. It took her less than five minutes to make a diagnosis and suggest a cure.

"Wolfie gets anxious when he's in the car," she said. "We don't know what his history is. Maybe his former owners dropped him off on the expressway. Now you have to teach him that the car is a safe place."

"How do I do that?"

"Simple," said Ellen. "Calm his body, and his mind will automatically follow suit."

She showed me how to get Wolfie to lie down and stay down after getting into the car.

"When he is lying down, he cannot get excited. This should do it."

Such a simple solution to this entrenched problem seemed almost silly. Why would lying down make a difference?

But it did the trick. Wolfie has never again gotten sick in the car. He sits calmly. Within Wolfie, and within my grandfather, lies the answer—a calm body is a calm mind. Not the other way around, as most people believe. By coaxing your body into a state of calm, you quiet your rational brain's internal mental chatter and allow the parasympathetic system and core brain to do what they do best: calm you down. Equanimity can emerge not only as a result of therapy funneling down from our higher centers to our body but also from the upside-down flood of impulses from our body to our higher centers. The "bottom-up" version of body over mind is where the answer lies.

Freud may have been onto something in advocating free association while lying on a couch in a quiet room. Maybe that act itself rids people of anxiety. If you did that every morning for forty-five minutes, it would be hard not to have some of that mellow feeling drift into the rest of your day—just as my grandfather's early-morning rituals helped him stay calm as he dealt with the Byzantine bureaucracies of Indian commerce.

Our rational higher brain, with its Cartesian logic, may excel at figuring out how to build superconductors and teslas, but we need our core brain and our parasympathetically wired bodies to help us get a good night's sleep and to cope effectively with the onslaught of conflicting demands on our attention. In our relentless embrace of a fast-paced technology-saturated world, we have pandered to our higher brains, frontal lobes, and sympathetic systems, leaving our parasympathetic systems and core brains in the dust.

And therein lies the root of our contemporary epidemic of free-floating anxiety and inability to relax. But I contend that there is a solution. In this book, I explore the neurology and physiology of our body's innate "calm" mechanisms and show how to get them back up and running smoothly. In the pages that follow, I explain why achieving the state of focused awareness I'm calling "calm" is crucial, not just for happiness and success, but also for health. And I confront the biggest challenge: how to choreograph the battle between our brain systems so that it becomes a well-orchestrated dance. Only by accomplishing this can we hope to find the elusive balance that leads to what my grandfather knew so well: a peaceful state in a turbulent world.

CHAPTER 1

The Un-Calm, Modern Brain

Lisa, forty-eight, is a lissome blonde with a breezy manner. She smiles easily. When we meet, she appears poised and in control. I get no sense of any underlying tension. But I suspect it's there because Lisa has come to see me for only one reason: She is desperate for relief from crippling migraines. Headaches have been part of her life since high school, but in the last several months, she has been waking up with them every day. She is on multiple pain medications, including narcotics. Recently, her neck muscles have gone into spasm, causing her additional agony. Not surprisingly, she tells me that she is feeling more and more anxious.

"It's interesting," Lisa remarks. "I quit the advertising rat race a year ago because I couldn't stomach it anymore. The daily deadlines. The constant pressure. The need to always be available. The better I got, the higher up I got, the worse the pressure. It never eased up. By the time I left, I was pretty much running the company. The only trouble was, I was running myself into the ground, and my marriage along with it."

Finally, at the urging of her husband—"That man should be sainted for all he has put up with"—and her therapist, Lisa left her job to pursue what she always dreamed of doing: working with flowers. She started a floral bridal business, and with her connections, she was up and running in no time.

"Suddenly, I was surrounded by beauty, I literally was," she told me. "I was living my dream. I should be filled with happiness. Instead, here I am," and she smiled wryly, with a self-deprecating wave of her hand. "I can't sleep. I'm in constant pain. I need neck surgery. I'm on enough medications to put a racehorse into a coma, but it doesn't seem to be doing me much good. I'm a complete mess."

Lisa retreated from a glitzy career as a big-time rainmaker in advertising to follow her dream and was, by all accounts, wildly successful in this new endeavor. But here she is: on multiple medications, beset by pain and worry, even as she should have been having the time of her life. What is going on?

As Lisa told me her story, I could see that even though she was living her dream, there was something happening inside her that was generating both her migraines and her anxiety and preventing her from enjoying this achievement. In fact, I thought the cause of Lisa's problem was a very basic one. Over the years, she had trained herself to override her body's built-in relaxation mechanism, the parasympathetic nervous system. This created an imbalance with her alerting mechanism, the adrenaline-driven sympathetic nervous system. So when Lisa is dealing with a nervous bride and an overwrought mother who is unhappy with the proportion of lilies of the valley to white roses in the bridal bouquet, she shifts easily into high alert, tamping down her body's chronically weakened protective calming mechanisms.

Lisa had habitually overlooked the needs and desires of her primitive core brain, catering instead to the demands of her more evolved frontal lobes. Now, instead of savoring the glory of her beautiful centerpieces, instead of enjoying the happy smiles of brides on their wedding day, Lisa's frontal lobes are off to the next wedding, the next bouquet, the next stop on the train schedule. In the process, she continues to maintain a skewed, top-heavy brain, a brain geared to hundred-meter sprints rather than the slower marathon that is a life well lived. Her frontal-lobe circuits, concerned with planning, judgment, and self-criticism, have been preferentially strengthened, and her more primitive core-brain circuits, which deal with relaxation, have become undermined.

In the short term, this diligence helped propel Lisa up the career ladder at the advertising agency, where she worked relentless twelve-hour days, and now that same conscientiousness lets her prosper in the flower business. It allows her to navigate an abundance of tasks and demands: Even as she obsessively reads her e-mails and texts, she pacifies a distraught bride, negotiates with a wholesaler, and orchestrates a dinner party for her son's graduation. But what is the cost of this frenzy of activity?

Lisa's need to pack more into every hour makes the true enjoyment of any one experience difficult. She's not alone. Multitasking has become the modern norm, not the outlier. Here's one scenario: I am eating lunch, sitting at the table. Period. Here's the more likely scenario: I am eating a sandwich while watching television and cleaning the kitchen, talking on the phone, and checking my e-mail. The joy of simply eating a meal, just tasting it, is rarer than it used to be.

This need to stuff every day with activity and stretch our waking time as long as possible—Lisa rarely enjoys more than five hours of sleep—is a peculiarly frontal-lobe creation. The frontal lobe is the high priest serving the temple of the virtues of industry. In the frontal-lobe-driven process of creating a false sense of urgency and viewing life as a race to the finish, the slow, intense cadence of the core-brain drumbeat is lost and with it some of our innate capacity for calm.

Interestingly, multitasking is something the frontal lobes are specialized to do. They are like the air traffic controllers at a busy airport, telling which planes to land when and on which runways. The frontal lobes are able to process immense amounts of information instantaneously, allowing smooth, seamless transitions in dealing with the staccato taskmaster that is urban life.

Lisa is a champion multitasker. But by creating an unsustainable internal disequilibrium between the conflicting needs of her core brain and her frontal lobes, between her relaxing parasympathetic system and alerting sympathetic system, she is slowly destroying her health. I knew that medication would not be the best way to treat Lisa's anxiety, migraines, insomnia, and chronic pain in her current state of constant stress. Aspirin and surgery may ease her back pain, Prozac may reduce her anxiety, Valium may help her sleep, but none of these would address the *cause* of all this internal turbulence, the underlying physiological dearth of calm. To rid Lisa of her symptoms, I would have to correct the chronic imbalance between the calming and alerting forces within her body and brain. Later, I discuss what I did with Lisa, a simple solution that got her back to health and off her crippling dependence on a plethora of increasingly ineffective medications.

Lisa may be an extreme example, but there is no doubt that this disturbing trend of chronic, learned inability to self-soothe has become an epidemic. Modern demands on our brains—from the insistent buzz of technology to the persistent illusion of urgency—dictate this ill-advised shift toward the alarm zone of our body and brain apparatus, at great cost to our well-being and peace of mind. The use of prescription sleep aids, narcotic painkillers, anti-anxiety drugs, and antidepressants has skyrocketed, tripling in the last decade alone. But drugs have not solved the problem. More people than ever complain of insomnia and anxiety. Depression is on the rise, as are attendant illnesses such as obesity, hypertension, and cardiac disease. Depression and anxiety are five times more common among today's *children* than just fifty years ago.

The problem lies in our "top-down" approach to treating the ailments of modern angst. In fact, from a brain or neuroscience perspective, a "bottom-up" approach may be more effective. The top-down approach assumes that our state of mind is dictated by our consciousness and rational thought, driven by the frontal lobes. In contrast, the bottom-up approach is an unconscious, body- and core-brain-driven road to our mindset. When a mother says to her son, "Tony, would you please calm down," she is using a top-down approach to calming her child's mind. She is appealing to the reasoning capacity of his frontal lobes and asking him to use this rational, conscious process to quiet down. Psychotherapy is also a top-down approach in that it uses language and reasoning, both frontal-lobe functions, to explore the reasons for the lack of calm. On the other hand, if a mother says to a screaming child, "Time out!" and sits him in a chair facing a wall, that's using a bottom-up approach, that is, quieting his body and having him slow down in order to achieve a sense of calm.

Alcohol and other drugs "calm" the brain by dulling our senses, working on all areas of the brain, on both conscious and unconscious perceptions, and are effectively both top down and bottom up. Meditation and even Freud's analytic couch attempt to use a bottom-up body-to-brain approach, but, as we will see, they are not powerful enough, especially when faced with the constant onslaught of daily calm busters.

Let's take a look at the workings of Lisa's brain and body to see how she got where she is, surrounded by flowers but barely able to stop and smell them. Just behind her pretty brow, housed in the hard casing of her skull, the largest parts of her brain—the frontal lobes—are hard at work, preventing Lisa from relaxing. They are instructing her that it is more important to be hypervigilant than to seek serenity. While all of us have frontal lobes that have evolved over the millennia to perform these functions, modern life has left us unable to balance this vigilance with the relaxation that is also wired into our brains.

Lisa has taken this natural tendency to the next level. So when she tries to relax, her frontal lobes send electrical signals down to her core brain, where the alerting sympathetic and relaxing parasympathetic systems interact, amping up her sympathetic system and turning down her parasympathetic system. From the core brain, these signals flood her sympathetic system, where Lisa's body is constantly preparing for a virtual Armageddon. Levels of the stress hormone cortisol run high in her bloodstream, her heart is beating a little faster than it needs to, and her breathing is a bit shallower and faster than it ought to be.

Lisa's frontal lobes and the billions of nerve cells that

form them are using all their power to keep her from harm. In her case, of course, "harm" means the loss of an account or failing to deliver the wedding flowers on time—hardly a looming disaster. If Lisa had been alive ten thousand years ago, the harm might have been a woolly mammoth, a genuine threat to her safety. Today, surrounded by dancing daffodils, the danger is more a sense of foreboding, a feeling that if she ever slowed down or stepped off the treadmill for even a little time, she would lose ground. But Lisa's core brain, evolved from her ancestors over millennia, cannot make the distinction between real and imagined threats. It reacts strongly, instinctively, to keep Lisa safe.

As I listened carefully to Lisa's story and examined her, I became convinced—from the perspective of the bottom-up approach—that unless we were able to calm Lisa's body, it was going to be difficult to calm her mind.

Lisa had longed to work among flowers all her life. Over the years, Lisa's hardworking frontal lobes enabled her to achieve phenomenal success in the advertising world, to put away the sizable nest egg that gave her the financial freedom to indulge in her dream. Her frontal lobes allowed Lisa to flourish in a challenging work environment, helping her make good, disciplined decisions, her eyes firmly fixed on yet another, more challenging goal.

But the price tag for these achievements was high. Excruciating migraine headaches. Nights staring at the ceiling, praying for sleep. Anxious days popping Prozac and painkillers. But now, here she is, not even a year into her escape from her old high-pressure life, and she is already back in the jail cell of anxiety. Because of the way Lisa has molded her brain, she cannot relax. Her long-preferred frontal lobes and her sympathetic nervous system will not let her unwind, to unplug on demand. Her parasympathetic nervous system, historically overlooked, is no longer able to take a stand.

This imbalance between regions of her brain had not always been present. Lisa recalled that when she was little, she could spend hours squishing the sand between her chubby toddler toes, watching it run up and through them. Summer was summer, with sand in sandwiches that somehow didn't detract from their deliciousness. She would lie for hours on a warm beach, watching the clouds float by and closing her eyes when the sun peeked out. As a toddler, she got tactile pleasure from the sand between her toes, and her frontal lobes did not veto her sensual idling. This "time waster" met her core-brain needs and allowed her parasympathetic nervous system to be in charge, to transport her to a state of calm. Once, when she was six, Lisa tells me, she sat so still and so long in the garden that she saw a flower open before her very eyes!

"I can't believe that I could sit that still for so long!" Lisa exclaimed in wonder as she recalled this moment. "You'd have to knock me unconscious now to get me to lie still. I always have to be on the move."

When Lisa was young, her frontal lobes were still being formed. They were not as powerful as they are now; her parasympathetic system still had the ability to lull her into a deep sleep. In fact, as is true for all of us, Lisa's frontal lobes continued to develop until she was nearly twenty-five. They had not yet been fully shaped by her personality, her family, and the world around her. Lisa was born with the ability to relax completely. As a baby, she cried when she got hungry, but sometimes her mother wasn't around to feed her. So she sucked her thumb to relax and the effect was calming. She fell asleep.

Relaxation is not the same as happiness, but it is difficult to find consistent contentment in life without being able to soothe oneself. All of us are born with this ability; we lose it to larger or smaller degrees as we grow up. When Lisa was young, her brain did not interpret playing, idling, or just relaxing as threatening or dangerous. But over time, she learned that sitting in the sand, watching the waves race each other to the shore, was not a good thing to do. What was good was purposeful activity, like setting the table or ironing or doing homework. Relaxation was fine, but only on vacation, on July Fourth, or at a barbecue.

"The idle mind is the devil's workshop," her mother would say when she saw Lisa swinging in the backyard hammock, watching the clouds float by. "What are you doing just lying there? Go read a book, clean your room, *do* something!"

"Doing something"—in fact, many things—was highly prized in young Lisa's home. As she grew up, she taught herself to multitask and soon became submerged in the culture of doing. Her frontal lobes internalized society's notions about the virtues of industriousness, and these values thrived in the fertile soil of Lisa's brain. The frontal lobes are also the seat of rational judgment, the part of the brain that *logically* determines if something is inherently good or bad, as opposed to the core brain, which *instinctively* makes these determinations. So Lisa's frontal lobes began telling her, via electrical signals traveling through her nerve cells, that relaxing was "bad," in contradiction to the message from her core brain, which regards relaxing as "good." Soon she lost the ability to daydream and to sink into a deep and healing sleep at night.

When Lisa tried to relax, her frontal lobes would shift into high gear, imagining all the bad things that could make her vulnerable to predators, whether the woolly mammoth or the creditor at the door. The frontal lobes decide what we consciously pay attention to and for how long; the core brain operates behind the scenes, unconsciously directing our attention. Our frontal lobes prevent distraction, help us resist temptation, and rack up more and more accomplishments. Lisa's frontal lobes are interested in her financial viability and her next customer, while her core brain is exclaiming over the velvet texture of the red rose.

"When I relax," Lisa confessed, "I feel guilty. I'm a workaholic. I need to always be busy. I don't want to be seen as lazy. When I was at the agency, if I wasn't available to my colleagues or my clients or, God forbid, my boss, it was seen as irresponsible. So I never let that happen."

Wisdom Overheard at the Airport

This fear plagues many successful people. Indoctrinated in the "do" culture, they worry they'll be considered irresponsible or lazy if they are not ready for action 24/7. Thinking this way fosters an atmosphere of unremitting watchfulness, a constant failure to calm down and self-soothe. I am guilty of this attitude as much as the next person. Not long ago, I was stuck at the Brussels airport on the way home from a conference. I was idling, waiting for my flight to board, and couldn't help overhearing a conversation. A slightly overweight and balding middle-aged man in shorts and sandals was talking to a younger man with long hair, sporting a black T-shirt, and carrying a guitar. Soon enough, the unlikely pair decided that they liked each other enough to exchange contact information.

"Here's my e-mail," I heard the musician say. "Let's grab a drink when we get back."

"Okay," replied the older man, "but you won't hear from me for a month. I have this strict rule—no e-mailing or cell phone when I'm away."

"Awesome, dude!" said the musician, giving the no-phone man a high five. "How do you get away with that?"

I was slouched in my seat, eavesdropping in a bored, distracted way. But now I sat up, echoing the musician's sentiments. "What was he talking about?" I said to myself. "How in the world did he get away with it?"

The older man spoke. "I saw a few years back that I was spending as much time dealing with work on vacation as when I was at my office. In finance, everybody wants their answers yesterday, their problems solved immediately. I couldn't get away even when I tried. My BlackBerry rang all the time; people got upset when I didn't call back or respond to e-mails right away. So I started sending out e-mails about a month before my vacation, letting people know I was going away to a place where I could not check e-mails or have cell phone access."

"What if someone needed to reach you? What if it was an emergency?" the musician asked, echoing my own unasked questions.

"It was hard for me at first," the man confessed. "I worried about the same thing. But then I thought, what if I died? What if I got sick? No one is indispensable. And neither was I. There was a lot of resistance at first. My assistant went into a panic. She kept asking me what she would do if a decision had to be made or if she needed information that no one had but me. But I trained a bunch of folks to be in charge while I am away. We have everything documented for what-if scenarios."

The musician looked longingly at the consultant. "Man, that's awesome," he repeated. "It would never fly with my friends, though, even if I wanted to do it."

"It's a reeducation process," said the balding corporate sage. "You've got to teach yourself and everyone around you that you are dispensable. People *can* manage without you. You have to believe it. Everyone is dispensable. Now, maybe I am not as high up in the corporate chain as I could be. But I am healthy, at peace. It's a no-brainer, if you ask me."

What amazed me about this conversation were not the wonderful truths in it. Those came to me later. What amazes me even now was my immediate reaction. I remember thinking, "What a shirker this guy is! I feel sorry for the people he works with, that they can't rely on him." My frontal lobes were quick to make these negative judgments about a stranger who, not unreasonably, wanted some real downtime. As I boarded the plane, I found myself wondering if I could pull away like that. Leave my office for a month. Disconnect from the Internet. No way, I thought. Then I reminded myself how much the core brain craves downtime, even as the frontal lobe gorges itself on multitasking. And I said to myself, "Hmmm, maybe. What if?"

A Sea of Anxieties

It is no surprise that so many of us are reluctant to unhook ourselves from the work tether, unable to escape the tentacles of technology. Whether we're in the tense confines of the corporate boardroom or trekking in the wilderness, whether we're out on a date or asleep in our beds, our cell phones and BlackBerrys are at the ready, keeping us addictively connected. As Lisa complained about her life in advertising, we are expected to constantly be available to others. One yank of our electronic leash and we are transported far away from where we are physically. Smartphones beep and flash and blink, sending sudden staccato signals that disturb the core brain's sense of calm even as they reinforce the pseudo-urgent demands of the frontal lobe.

Constant attentiveness is one reason we are increasingly powerless to self-soothe and relax. Our core brain, as well as our more sophisticated frontal lobes, are programmed by evolution to dread the "threat of disconnection." In our ancestors' time, when danger lurked all around, separating from the protection of family or tribe could mean sudden death.

Fear of disconnection is only one of many modern deterrents to self-soothing. Consider, too, the strain imposed by the myriad roles each of us must play in our increasingly complex society. Men today are expected to be devoted fathers and soccer coaches, husbands and lovers, workers and friends. Even as women have soared to the top in the workplace, they still shoulder the bulk of family responsibilities and are more likely than ever before to be unhappy. Both sexes are rattled by new uncertainties in social norms. Does the woman split the dinner check? Does the man hold open the door? Do we comfortably accept the idea of husbands who run households and wives who run companies? Society demands that we play all these roles well, and in an age of instant communication and information overload, the bar on what constitutes "well" keeps going up. The frontal lobes juggle these more complex negotiations among our many conflicting roles.

Electronic technology itself is a culprit in the struggle to stay calm. Not only does the cacophony of signals that assail us day and night overwhelm our nervous systems, but our high-tech environment also disconnects us from real-time, intimate, undistracted day-to-day dealings with nature and with other people. And these connections are crucial for the core brain, which is wired for connections with other humans, particularly by integrating our senses with our responses to others in our tribe. Twitter and Facebook enable us to form virtual communities all over the world—expanding our knowledge and enlarging our social circles—but they do not satisfy the core brain. This unfettered access to our virtual friends may help our frontal lobes feel connected, but the core brain needs touch, sight, smell, sound, and three-dimensional here and nowness, and it is left unsatisfied by a Skype rendition of a far-off loved one.

When Lisa starts checking her BlackBerry, communicating with her next customer, she may assuage her compulsive, forward-thinking frontal lobes, but she denies her core brain the squishy joy of watching the bride delighting over her bouquet. She misses out on experiencing the positive emotions she has brought another human being. In this way, Lisa's brain is incessantly negotiating with itself, balancing future possibilities with current pleasures. A calm, serene, healthy Lisa results when her brain regions work together in harmony rather than continually overriding each other.

Contemporary culture mounts still another assault on our capacity for calm. We are expected to have infinite knowledge about a multitude of things, much of it abstract and conceptual. Lisa frets when she cannot remember the names of each and every customer she deals with. But her brain, and yours and mine, is wired to remember tangible events in the real world, like the smell of garbage or the sound of someone's voice, not a catalog of facts and figures. Our brains, says Michael Gazzaniga, a leading neuroscientist, "were not built to remember the kinds of things we must learn in the modern world . . . the brain is built for organic things, such as remembering where real harm can come to you in real physical space."

Lisa is far more likely to remember the time when a little flower girl ate all the rose petals in her bouquet, and how everyone laughed, than the names of the people in the wedding party. Our brains are not wired to remember dates and telephone numbers but rather how to avoid getting hurt and the physiognomy of a beloved face, the memory of which the core brain sears into our brain. Personally relevant stories and a face convulsed with pleasure activate the core brain, making these events indelible memories.

Today's emphasis on retaining and recalling details, which is largely a frontal-lobe skill, flies in the face of a brain wired to remember thematically. In preliterate cultures, certain select individuals, called griots in West Africa, were the repositories of the oral history of the tribe and were chosen precisely because they had better memory skills than their fellow villagers. Now every person is required to remember vast amounts of information for which the brain is not wired. Not remembering and the fear of forgetting creates a tremendous amount of stress.

Another potent source of stress is the dazzling variety of choices we face every day. When you have too many choices—whether you're picking a spouse or selecting a salad dressing—the frontal lobes become less efficient. They process the pros and cons of each choice but veto the core brain's input. When the core brain goes "Oooh!" at the sight of a salad dressing (or a prospective mate), the frontal lobes may respond with "too many calories!" or "not well connected enough." When there is a chronic mismatch between frontal logical needs and core-brain instinctual needs, stress as we know it, of both body and brain, begins.

Indeed, in many ways, choice is the enemy of calm. The more mobility we have, the more career possibilities available, the larger the pool of life partners and sneaker brands we can choose from, the more anxious we become. Our older, core brain, never good with nuance, gets lost among the twenty different types of extra-chunky spaghetti sauce. The complex, nuanced decision-making process that is frontal-lobe choice gets in the way of the simple, instinctual decision making of the core brain.

Consider online dating, where we can specify all manner

of requirements in a prospective mate, from placement of tattoos to love of Kierkegaard. With this amplified sense of choice comes the terror of making the wrong selection. This is another scenario in which the frontal lobes are doing the picking when, in fact, the core brain should be in charge. The core brain needs real interaction with real people to assess risk and make decisions. Our rational frontal lobes are more prone to self-deluding acts, using logic to overcome instinctive core-brain unease. Even as your core brain picks up on the slightly shifty eyes and change in body orientation and odor of your partner when you ask her whether she has been faithful while you were away on that business trip, your rational frontal lobes register that she has called you daily and said she missed you. Of course she is being truthful. The core brain is an excellent "bullshit detector" and is far better than the frontal lobes at knowing whom to trust.

Trust in others, particularly in urban regions, is one more uniquely modern deterrent to calm. City life presents a daily conundrum of whom to trust and how deeply to trust. Our core brain is on the alert for signs of deceit or danger in those we mingle with, but it is often overridden by our frontal lobes, which tell us there's nothing to fear. People of different ages, genders, religions, colors, and customs populate our modern cities. We have no history of knowing people over time; interactions are typically quick, one-time deals made with no past social history.

The core brain and our frontal lobes try frantically to

sort out the "good" from the "not good," but this is difficult to do. We evolved from a time when we knew between fifty and a hundred and fifty people in our lifetime to an urban culture in which we may encounter that many new people in an hour. The human brain is built to react not just to itself and the environment but to other people. Meeting this many people on a daily basis fosters a certain level of vigilance, a lack of trust, which puts your core-brain sentry on alert, even though your rational frontal lobes tell you to relax. How do you address the core brain's instinctive fear of trusting a stranger while listening to the logical litany of the frontal lobes, which assures you that the man in the cop's uniform is trustworthy? The choice comes at a price: anxiety. Urban city living is chock-full of such situations, in which ancient core-brain fears are overridden by frontal-lobe logic.

Closely related to the problem of trust as an impediment to calm is the conflict between altruism and self-interest. Even as modern society values altruism, a frontal-lobe value, it rewards self-interest, a core-brain characteristic. However, if we follow our core-brain dictates, we worry about society's recriminations for our "selfishness." This juggling between common good and selfish good causes unease and anxiety. It is noteworthy, too, that the core brain finds comfort in hierarchy even as the frontal lobes champion democracy. Derived as it is from lower mammals and responsive to hierarchies in families, the core brain does not do as well as the more evolved frontal lobes with the choices of equality and democracy. Similarly, the core brain stumbles when it chances upon abstract concepts like nepotism.

Nepotism—favoring our own family—is frowned on by the frontal lobes but is wired into our core brain. Nepotism, essentially tribe-speak, allowed for survival in ancient times. As elegantly argued by Richard Dawkins, our genes are "selfish," interested primarily in their own survival and concerned about others only insofar as their survival might have an impact on ours. Self-interest is as hardwired in our core brains as in our genes, and overriding this basic need is a deterrent to calm, for the core brain perceives such measures as threats. We are more likely to help those who will advance our interests. We're more likely to save our own children than a neighbor's from a burning house. Would you use your connections to gain admission for your daughter in the gifted program or to get your brother a job? Even if it means someone else's more gifted child will not get into the program, even if it means that a far more qualified stranger will not get the job? If we are honest with ourselves, we would answer yes to these questions, although we overtly champion a meritocracy of equal opportunity.

In urban environments, we interact with thousands of people over our lifetimes, but we never build relationships with the majority of those we meet. In a small rural community, common good and personal good often have the same goal, but this is less often true in larger, more layered societies. Self-interested "what's in it for me?" behavior occurs when you have no commitment to the other person, when you have no future together.

Loss of intimacy is another modern hurdle, another deterrent to calm, as true intimacy, with its shared emotions, is a salve that delights the core brain. Being part of a group that binds in this real way allows us to face the world in better shape, more calm and less anxious in the face of trouble. This is one reason why people who follow sports teams are less likely to be depressed or anxious. Sharing the feelings of loss or wins with other humans helps us feel connected at a very basic level. A common problem for executives, celebrities, and professionals is the loneliness at the top. People in leadership positions are increasingly cut off from "village" connections, knowing who is doing what, when, and where.

Loss of personal closeness may be one reason for the perennial popularity of *People* magazine, which offers us an imagined intimacy with both celebrities and ordinary folk. My friend Lauren is a Yale-trained lawyer known for her superb technical expertise and her disdain of people she thinks are less intelligent than she. Her caustic comments to colleagues leave her mostly friendless and isolated. Viviana is my manicurist, a plump and pretty young woman who never graduated high school. With her ready smile and soft touch, she is well liked and popular at the salon. Ian runs a hedge fund and is a pleasant middle-aged man with a wife and three sons. He makes little jokes, has a self-deprecating manner, and is easy to be around. Then, there is me.

What do Lauren, Viviana, Ian, and I have in common? We all love reading *People* magazine. While Lauren, Ian, and I turn to it surreptitiously, when we're sure no one can see us, Vivi reads *People* openly and takes great pleasure in discussing the latest scandal by this B-list celebrity, that A-list actor. What is it about this glossy chronicle of the adventures of the famous and not-so-famous that appeals to each of us, disparate as we are? It has to do with the village mentality of the human brain, which is community oriented. It hungers for information about people, the real lowdown, the gossip. In an increasingly segmented and fragmented world, the continuity provided by "knowing" these strangers may be soothing to the core brain. It provides a measure of calm.

People's popularity speaks to the need for community. Even as the frontal lobes champion the acquiring of possessions and keeping up with the Joneses in a McMansion-studded gated community, the core brain seeks closeness with the Joneses in a yurt. Physical intimacy trumps riches in core-brain values. Modern society, while addressing in painful redundancy many of our physical needs, fails to satisfy the core brain by giving short shrift to communal needs.

Fear in any form—whether of strangers, lost intimacy, or dangerous animals—is always a barrier to calm, even if only temporarily. Many such fears have their roots in evolution and are inherited in our genetic code. Most of us are instinctively afraid of snakes, while few of us are afraid of butterflies. Excessive, unexpected noise is not only an urban nuisance but also a primordial fear. The importance of detecting noise for survival is clearly in evidence during dream sleep. Your body is immobilized in this stage of sleep to prevent you from acting out your dreams and harming yourself. However, your eye muscles and inner ear muscles remain functional, allowing you to awaken should you hear any unexpected noise—a prehistoric life stressor. This ability is a great boon for the makers of alarm clocks, but every time we hear the clock's buzz or bell, we are actually being alarmed into awakening, which is really not a good way to start the day.

Fear of loud noises, of darkness, of heights, all helped our ancestors exercise extreme caution and thus survive to pass on their genes. In modern life, we are asked to override such fears, as with alarm clocks. How does this ignoring of core-brain alarms affect our ability to be calm, even as we have no conscious awareness of the noises that reverberate through our cities? Soon after the September 11 attack on the World Trade Center, I became sensitized to the sound of airplanes. Before 9/11, I barely noticed their steady drone overhead. But after that tragic day, I became consciously alert and sometimes panicked at the sound of a plane flying too low, a not uncommon occurrence in New York City. In time, my fears faded as my rational frontal lobes talked me into reacting less. But I believe my core brain is still on high alert when it hears the noise of an airplane.

Fear of heights and darkness are also carryovers from evolutionary times. Studies have shown that human babies and day-old goats will not walk on a transparent solid surface that gives the impression of heading off a cliff. Groups that do not have fear of heights, for example, the Mohawks who erected the Empire State Building, are said to inherit this ability. Yet, here we are constantly traveling by plane in a society that frowns on a fear of flying. We live in towering buildings, often with sheer glass faces. These perceived dangers trigger core-brain anxieties, overriding the rational frontal-lobe reassurances. Remember, the core brain is not logic driven. When you look down at the street from the terrace of your twentieth-floor apartment, your core brain is in mortal fear. Your frontal lobes may reassure the core brain by redirecting your focus to the thick glass that separates you and the long, deadly fall. But the core brain is still scared stiff.

We are hardwired to seek the company of other humans when we are afraid, even if it is to our detriment. When a little boy wants to cuddle with his parents because he is afraid of the dark, when a toddler won't enter a room until her mother turns on the light, evolutionary fears are at play. Denying children this comfort, even though their fears are not "realistic," may come at the price of stress and anxiety for both the parents and the child.

Toward an Alternate Reality

With so many hurdles to a calm state and so many triggers for unease, it is no wonder that so many of us are awash in anxiety. When our frontal lobes try to override core-brain fear—whether the ancient terror of heights and darkness or the modern reaction to the mandates of multitasking—we can expect physiological reactions. This is what may have led to Lisa's migraines and anxiety. Her resting heart rate rose, her blood vessels became more constricted, and her gut more sensitive. Her hypervigilant frontal lobes and hyperactive sympathetic system have helped further Lisa's career, but her body and brain have paid a stiff tax for the privilege.

There is an alternate reality. If Lisa reins in her bloated frontal lobes and hones her core-brain circuits so that her parasympathetic system is in balance with her sympathetic system, she will be relaxed and successful, well rested and fulfilled. She will have the same brain, but now it will be using a revamped program: It will be directing the brain's hardware by running new software. This book's premise is that we are all born with more or less the same brain hardware, but experience molds our programming. The upside is that all of us have the ability to achieve a state of calm with the right reprogramming. Later, I detail some of the traditional and more intriguing ways in which this reprogramming can be accomplished.

As we've seen, to understand the neurology of calm, we have to understand the workings of our core brain and our newer, rational frontal lobes. The body and the brain have perfected homeostasis, the painstakingly elegant process of balance, one that goes on from the smallest skin cell in your little finger to the large tortuosities of your colon. The purpose of this system, with its many checks, is to prevent imbalance, lopsidedness, of any sort. When we interfere with this homeostasis, when there is no balance between core brain and frontal lobes, the result is dis-ease.

We need to understand how our body is always on a seesaw, balancing calm with anxiety, balancing the right amount of alertness with the right amount of relaxation, the right amount of logic with the right amount of emotion. Good balance between the core brain and the frontal lobes, and between our parasympathetic and sympathetic nervous systems, creates calm and a sense of well-being. Imbalance creates anxiety and unease. How do we get to this balance? To answer that question, we must take a closer look at the core brain and its major instruments, the parasympathetic and sympathetic nervous systems. And we must investigate the little-understood vagus nerve, our body's chief instrument of calm. Together, the vagus nerve and the core brain hold the key to unlocking the secret of calm.

CHAPTER 2

The Core Brain, Architect of Calm

Walt was the third child of his mother and was born shortly after the end of World War II. He was a full-term baby, delivered at home in Los Angeles County after a normal labor. He cried lustily at birth.

Walt's mother brought him to the hospital when he was only six days old. At the hospital, the doctors noticed that, like any other infant, Walt cried when handled roughly and showed "contentment" when he was cuddled. He cried when he was hungry, sucked vigorously, and slept after feeding. If he was dipped suddenly as if being dropped, he would throw his arms out in fear. He firmly held on to a finger when it was inserted into his tiny hand. Little Walt lived eighty-five days before succumbing to his condition.

Walt sounds like any sick baby in the hospital, but in fact he had a disorder so grave that it was incompatible with life. Walt was born with only the rudiments of his core brain, the brain stem. Where the rest of his brain and his skull should have been, there was only skin and hair covering "a larger mass" that had "collapsed." He was blind because the nerves from his eyes had no brain to connect with. He was deaf for the same reason. He had no ability to feel with his skin, and his apparent comfort at being held was mediated through a system that did not involve the spinal cord.

His doctors concluded that his "cry of fear on rough handling" must have originated from Walt's primitive core brain. The instincts of sucking, sleeping, and waking, along with the "emotions of comfort and discomfort," must also be "patterned" in the core brain. He did not need a rational higher brain and frontal lobes to experience and express many emotions and instincts, including fear, contentment, and hunger.

Drs. J. M. Neilsen and R. P. Sedgwick presented their clinical and autopsy findings on Walt at the third annual meeting of the Society of Biological Psychiatry in Atlantic City on June 12, 1949, under the title "Instincts and Emotions in an Anencephalic Monster." I saw the photographs of Walt, and he is no monster. His large and sightless eyes gaze at us endearingly from half a century ago.

Since the publication of this presentation in the *Journal of Nervous and Mental Diseases* in November 1949, 47,491 scientific articles have cited this paper, the last citation a week before I researched the article. To put this figure in perspective, the average number of citations per scientific paper is 12.6.

The baby was called WLT in the paper, but I took some minor literary liberty in thinking of him as Walt, a little baby boy, amazing in his own way. Little Walt may have had a short and emotional life, but sixty years later, he shows us the power of the core brain and sparks new research and thinking about the core brain and the architecture of calm.

Walt taught us that emotions do not require higher-brain activity. With just a functioning core brain, he behaved like a normal infant, even though he could not see, hear, or even feel, in our conventional understanding of the term. Because Walt could not experience normal pain or sensation of his external body, he did not cry when his diaper was wet. He could experience only sensations coming from *within* his body, because these were carried up by the vagus nerve, part of a system separate from the spinal cord. Walt was unable to *think*. Even so, he managed to get hospital staff to hold him, interact with him, feed him, and care for him. Six decades later, he got me to smile and write protectively that he was "no monster." Walt had his core brain to thank for all this.

Because the core brain is geared toward survival, any process that interferes with our sense of well-being, our sense of "calm," is of necessity first perceived in, first processed, and first tended to by our core brain. This core brain reacts instantaneously to all environmental stimuli, responding without waiting for direction from higher brain levels.

Indeed, waiting for the slower higher brain centers to react might spell the difference between life and death. These core-brain sensors were what allowed little Walt to feel comfortable and content or afraid and fidgety.

Walt lacked even a rudimentary capacity for language, yet he was able to soothe himself. In contrast, Lisa, blessed with a perfectly good brain and an elegant command of language, could succeed in business but was unable to calm herself. How was Walt able to feel and then convey his feelings to those around him, to communicate and calm and soothe himself, even as he was missing nearly all of his brain? Why was Lisa so powerless to calm her anxiety?

The answers lie deep within Walt's and Lisa's skulls, in the core brain, the oldest and most powerful part of the brain. Our core brains are remarkably similar, in dogs and humans, in cats and apes, across species; they are the part of the brain that is most tightly hardwired to the body. Lisa ignored the needs of her core brain and was lost in the clamor of modern life, a marionette under the control of her rational frontal lobes. Working harder and harder, she expected her frontal lobes, her intelligence and diligence, to keep her from harm. But true freedom cannot be purchased by satisfying the frontal lobes, as Lisa discovered. Real calm comes with gratifying the core brain, which Walt was able to do. Every time Walt was cuddled or stroked, each time he suckled and filled his belly, his core-brain needs were satisfied and he became soothed and calm. If Lisa had taken the time to share in the joy she brought her brides and drown in the scent of her lilies of the valley, she, too, would have been gratifying the needs of her core brain.

A Trunk of Nerve Fibers

Physically, the core brain is a thick trunk of nerve fibers and clusters of nerve cells, about as thick as your thumb, which bridges the rest of the brain and the spinal cord. It is nestled between and below the larger lobes of the brain just above the nape of your neck. The core brain is the "automatic," instinctive, unconscious center of the brain; we are hardly ever aware of it. Neurologists refer to the core brain's lower, evolutionarily older part as the brain "stem," atop which flowers the newer, rational, thinking brain.

The core brain is responsible for Walt's and Lisa's breathing, their heart rates, the dryness in Lisa's throat, and the queasy feeling she gets when she's scared. The core brain is also the reason the doctors smiled when Walt cooed. The physicians' own core brains were wired to respond to his expression of feelings. When someone else's core brain is emoting, your core brain picks this up and relays it as a "real" emotion to your brain. This is why, when you want to be calm, it helps to be around calm people. In Walt's case, being able to be social was as crucial as the need to signal hunger and discomfort. The core brain is essential for surviving the elements and also for achieving social success, which in ancient times translated into survival.

The core brain is the part of the brain where calm and anxiety are integrated with the sympathetic and parasympathetic nervous systems, with rational and emotional thought, with the body and the brain. If there is no calm at this level, it is nearly impossible to achieve calm at a higher, rational level. In other words, if your core brain gets input from your body that suggests a state of high alarm—say, when you see a man who looks like the guy who mugged you two years ago—your core brain keeps you in that high-alarm state, even though your rational brain says, "He's not the mugger; he just looks like him." Your memory of that terrifying attack registers at the core-brain level and puts you on edge, making it hard for you to become calm. This dissonance between the reassuring messages of your rational higher brain and the messages of your primitive but powerful core brain lead to anxiety.

Although I speak of the core brain as an indivisible and singular entity, in truth it is a conglomeration of circuits. Parts of the core brain deal with blood pressure control, parts deal with gastric-juice secretion, parts deal with fear processing, parts respond to smiles, and parts get us to throw up when we ingest what the core brain perceives as poison. These core-brain circuits are the reason we "instinctively" like someone and "instinctively" dislike someone else. They are what convince us that a friend is lying even as he swears he's telling the truth. In short, the core brain is responsible for what we call our "gut feelings."

The Vagus Nerve

So what is the principal input to the core brain in this unconscious, emotional circuit? The vagus nerve is a key and often overlooked conduit of input that is important for the feeling of calm. It is a far-flung, meandering Nile of a nerve that ferries information to and from nearly all the body's organs and the core brain. For the most part, we do not consciously perceive the constant flow of information back and forth along the vagus, which keeps our blood pressure stable and our saliva flowing. This intricate vagal pathway is distinct from the parallel, consciously perceived spinal cord pathway through which our visual, tactile, auditory, and gustatory senses travel. Nerve endings dispersed in blood vessels throughout the body ferry information back to the core brain, where the vagus modulates such information. Given this level of reach, the vagal network, filamentous though it is, is just as important a "news" source about the state of the body as the better-known spinal cord system. Generally speaking, we are more conscious of information sent up to our brain through the spinal cord than information sent up to our brain through the vagus, which is often processed at the core-brain level.

The vagus nerve transfers information between our core

brain and our bodies, particularly our internal organs and aspects of our face—smiling, frowning, pouting—that convey emotions and are thus necessary for social survival. Good poker players become expert at controlling their core brain and vagus nerve so that when they have a good hand (as determined by their frontal lobes), their bodies and faces, controlled by the core brain via the vagus, don't betray them (and their cards).

The vagus is like the roots of a tree, hidden from conscious view, mining information from our inner organs and structures and taking them back to the core brain and eventually up to the frontal lobes. It is because of the vagal core-brain circuit that Walt was able to cry when he was hungry, smile when content, and communicate with his caregivers. Walt's vagus nerve fed back impulses from his organs to his core brain, where they connected with the nerves of his expressive, emotional face. Thus, the vagus and the core brain helped Walt convey his contentment and his distress. It was because of this circuit that he was able to survive for three months without a brain as we define it.

As far back as the 1800s, Charles Darwin recognized the importance of the vagus, declaring that "when we are disaffected it reacts on the brain; and the state of the brain again reacts through the vagus nerve on the heart; so that under any excitement there will be much mutual action and reaction between these, the two most important organs of the body." Darwin wrote of the value of the vagus in maintaining our emotions and pointed to it as the primary conveyer of information between our brains and our hearts. Nearly two centuries ago, Darwin realized how vital the vagus and the core brain are in maintaining a state of calm.

A sense of community, of shared empathy, also plays a crucial role in achieving calm. We see this in baboons and dogs and other social animals, who soothe themselves with grooming and communal activities. The core brain has multiple features that allow us humans to flourish in communities and thus to live in an environment that can foster empathy and calm. Chief among these features is the ability to quickly and accurately read and respond to the emotions of others. To facilitate this, the core brain contains large clusters of nerve cells called pattern generators. When you start to cry because you are sad, a specific pattern gets activated: Your face scrunches, you begin to tear up, your breathing changes. When you are afraid, your eyes widen and your pupils dilate, your heart rate increases, your sphincters change tone, your gut slows down. These universally recognizable patterns of bodily response to emotions are due to the webs of networked nerve cells in the core brain.

The Core Brain System

According to Harvard neuroscientist Clifford Saper, who introduced the concept of pattern generators, these core-brain responses evoke movements of intricately linked muscles to convey *true* emotion. Notably, though, these movements are not invoked by false attempts at emotion. When we laugh at a joke because we find it funny, the pattern generator for genuine laughter is evoked, the muscles around our eyes crinkle, our mouth widens, and we emit sounds from our throat, in a choreographed near-instantaneous sequence of events.

The friend who told you the joke (and especially her core brain) realizes your laughter is genuine. But when you laugh at a joke that you actually don't find funny, and are just trying to be polite, the pattern generator of laughter is not evoked and the laugh is stilted, the choreography of muscle movements is off. (This is why saying "cheese" when posing for the camera leads to an artificial "cheesy" smile.) Your friend's core brain immediately recognizes your laughter as false, but her rational frontal lobes may override the core-brain alarm. This is because, unlike her core brain, your friend's frontal lobes are swayed by other motivations. She may think the joke is funny and not really notice that you laughed just politely. She may be an aspiring comedian at an audition, looking for positive feedback. At any rate, her frontal lobes are easily fooled by the false display of emotions, but her core brain is not. This is one reason why face-to-face interactions are preferable for business negotiations and for falling in love. It is harder to deceive the core brain.

Disgust, anger, and surprise all involve pattern generation of groups of muscles and changes in heart rate, bowel irritability, and breathing. Good actors recruit these pattern generators to convey emotion, tapping into their own memories and sensations to convey sadness, for example. If they don't do this, the acting falls flat and fails to generate feelings within you.

Interestingly, many animals, including cats and dogs, have these pattern generators in their core brains. So when a person "snarls like a dog," he really is snarling like a dog, because similar sets of pattern generators are invoked in the core brains of both the dog and the person. Because of the pattern generator system, it is very difficult to move a muscle conveying emotion in isolation and requires much practice, as anyone who's tried to voluntarily flare their nostrils knows.

It is notable that only higher apes and humans have the ability to be deceitful in displaying emotions. We are able to show our teeth as if in a smile without really experiencing the emotion of happiness. Lower animals cannot do so. What they feel is what we see. What we feel is what they sense. This may be why dogs "instinctively" know when someone is afraid. The dog's core brain is picking up the signals that your core-brain fear pattern generator is giving off, even as your rational frontal lobes say, "Look! It's only a Chihuahua. Nothing to be afraid of!" But the Chihuahua *knows* you are afraid. If it is not a particularly nice Chihuahua, it advances, it snarls, and before you know it, all 210 pounds, 6 feet 1 inches of you is cornered and held at bay by 5 pounds of dog and teeth. You have only your core brain to blame for this.

Understanding the part the vagus and core brain play in the production of calm illuminates why so many of our attempts at achieving calm may fail. We place so much emphasis on what we consciously see, what we consciously feel, and what we're able to process with our rational brain that we often ignore the far stronger processes of our core brain. Rarely do core-brain perceptions reach the realm of consciousness, and even then, our frontal lobes often rationalize away our fears and gut instincts.

We agree to sign the apartment lease, even as our core brain "advises" against it, because our frontal lobes have always wanted an apartment with a terrace. So we choose to ignore the barely perceptible changes on the broker's face when he says the cracks in the masonry are not structural. Our core brain sends up a red flag, but our frontal lobes talk us out of our apprehension. And such dissonance between core-brain and frontal-lobe impressions leaves us anxious. Anxiety and calm, along with happiness, sadness, and fear, are all emotions evoked by our core-brain pattern generators. In modern life, these pattern generators are handicapped by myriad "disconnectors" from our fellow humans, including technology that substitutes for face-to-face communication. This lack of tangible information discombobulates the core brain and confuses the vagus, giving the nuanced, choice-driven, rational frontal lobes the upper hand, making us vulnerable to anxiety. This type of endemic conflict creates a chronic, low-level sense of anxiety and a lack of calm in our systems.

Why is our brain so hardwired for the detection and expression of emotions like anxiety? Why is emotion relegated to the same part of the brain as breathing and blood circulation? What is it about emotion that makes it so important? Is it possible—or desirable—to exist without any emotion whatsoever? What would such a person look like?

Meet the hexamethonium man. As envisioned a half century ago by the pharmacist W. D. M. Paton, the hexamethonium man is "a pink complexioned person. . . . His handshake is warm and dry. . . . He is a placid and relaxed companion; for instance he may laugh, but he can't cry because the tears cannot come. Your rudest story will not make him blush, and the most unpleasant circumstances will fail to make him pale. His socks and his collars stay very clean and sweet (he does not sweat). . . . But his health is good. . . . Diseases of modern civilization, hypertension and peptic ulcers, pass him by. . . . He is thin because his appetite is modest; he never feels hunger pains and his stomach never rumbles. . . ."

This hypothetical creature has no parasympathetic or sympathetic control of his organs. Hexamethonium is a chemical that blocks the activity of these systems. The hexamethonium man is not hungry, has no discernible emotions, and is not plagued by diseases that are stress related. But our dry hexamethonium man seems so very boring, so much more of a "monster" than our more emotional Walt. Unable to perceive and express emotion, he would be a far cry from a desirable companion. Even in the unlikely event that he were able to survive the physical limitations of his condition, his genes will not live on, for his lack of emotion will prevent him from finding a mate. A man without emotions will simply not survive.

Evolutionary biologists know emotions are conserved across species because they are important for mating and for survival. Emotions are housed in the same parts of the brain as those that control our bodily functions, because without emotions, you are destined for evolutionary death, just as without a beating heart, you will not survive. It is clear that we are not meant to exist without emotions, no matter how desirable that state may sometimes seem. We survive emotionally as a result of the delicate balance between the fright-fight-flight sympathetic system, with its familiar adrenaline rush, and the less familiar calming parasympathetic system. The sympathetic system is the one that keeps us alive by getting us out of rough spots.

You are walking down a dark street in high heels because your car broke down. Suddenly, out of the corner of your eye, you see two unsavory characters bearing down on you. Your core brain activates your adrenaline-driven sympathetic system before your frontal lobes even register that the two fellows are up to no good. You are running before you *know* it. (This is not simply an expression, by the way. Until it reaches and registers in your conscious higher brain, your core-brain activity is not known to you.) Your heart pounding, you outrun your pursuers to safety and then discover, to your amazement, that you ran three miles with a broken heel.

At home, you fall sobbing into your husband's arms and allow yourself to be soothed with a hot bath and a chilled glass of wine. The parasympathetic calming system restores your sense of calm and helps you feel safe in the arms of your loved one, just as the sympathetic panic system enabled you to flee your predators. The parasympathetic system slows your heart rate down, makes you breathe slowly and deeply, and fills you with a sense of contentment.

The sympathetic system allows you to fight and win battles, to get aroused and excited. The parasympathetic system allows you to relax and to fall in love. After a period of protracted sympathetic overdrive, for example, after a battle, it's crucial to calm down and relax. Such postexcitatory relaxation of the sympathetic system increases parasympathetic activity and helps prevent posttraumatic stress disorder in those exposed to severe trauma.

Interestingly, in days of yore, warriors engaged in actual high-sympathetic combat for only hours or at most a few days. Following a battle, they would indulge in high-parasympathetic restorative celebrations—feasting, drinking, and making love. Modern-day combat forces soldiers to remain in high-stress, high-sympathetic states for a year at a time, without true relaxation or restorative parasympathetic calming activity. There is data to suggest that current high posttraumatic stress disorder rates may be attributable to the protracted periods of predominantly sympathetic energizing activity without any intervening restorative calming parasympathetic activity.

Posttraumatic stress disorder is an extreme version of anxiety and stress. But, as we've seen, lack of calm in less acute form is an everyday feature of modern life. We know that calm is created by a state of equilibrium between the two systems, parasympathetic and sympathetic, and that anxiety and stress result when the alerting sympathetic system and rational frontal lobes override the core-brain and parasympathetic system, with the vagus nerve as its instrument. The big question, then, is: How can we get our parasympathetic system and our vagus nerve to be more in charge so we can relax? The answer is both simpler and more complex than you might think.

CHAPTER 3

Getting to Calm from the Bottom Up

Annie is thirty-five years old, petite and pretty. A securities trader with one husband, three kids, and four long-haired dachshunds, Annie manages never to let her mascara run even as she trades with the precision of a fine Swiss watch. I have been following Annie for many years for a benign growth in her brain that has remained stable and has not required surgery.

At Annie's last six-month checkup, she told me she was worried about money, even though she was earning well into seven figures. She has seen friends and colleagues roiled in the recent fiscal turmoil, and now she is conjuring up all kinds of calamities that might befall her and jeopardize her family's future.

"What if my tumor suddenly starts to grow and kills me while my kids are at school?" Even for Annie, who is a first-class worrier, this was extreme.

"Annie," I scolded her. "You know that your tumor is extremely slow growing and benign. We would catch it well before it became an issue."

"I know I'm being ridiculous," Annie admitted. "But lately, I'm turning into Chicken Little. The sky always seems about to fall."

I listened quietly, not wanting to stir up Annie's anxiety. She talked about the economic downturn and how her own losses, though not catastrophic, were making her nervous. Her father had lost his savings in a risky endeavor when Annie was fifteen, bringing about wrenching changes in the family, which had led to her father's early death from a heart attack.

"I don't want to end up like my dad," Annie said. She had recently taken up yoga to try to reduce her anxiety. "My husband said yoga would be good for me because I've become so hyper. So I said to myself, Ed is right, this will calm me down. Then I started taking a class and I feel stupid because it's not having any effect. I don't easily recognize the feelings of relaxation. When the teacher says empty your mind, I think, what do you mean, *empty* your mind?! It's ridiculous.

"I overthink everything. The worst is when the guy says, 'Imagine yourself floating.' I worry, what if I drown, what if there's a tsunami? I'm most calm when I'm in the shoulder stand, upside down; then my brain really goes blank. I could even sort of take a nap in this position."

When her yoga teacher urges her to empty her mind,

Annie's frontal lobes respond rationally. "What's he talking about?" she says to herself impatiently. But when she stands on her head, her core-brain-driven parasympathetic system takes over; to her surprise, now she is able to "go blank." Annie's experience demonstrates how the vagus can create calm from the bottom up—literally with her bottom up—and override Annie's frenetic frontal lobes, to get her to "empty" her mind, to calm down.

How does this happen? What is it about the shoulder stand that makes it possible for Annie to empty her mind? Answering this important question means looking first at how Annie's circulatory system responds to different bodily positions. When she is in an upside-down position, there is more blood available to fill Annie's heart. As her legs and the rest of her body assume a less dependent position, more blood is moved into circulation. As more blood moves into circulation from her tissues, it fills her heart chambers and her heartbeat slows down; with each beat, it is able to pump more blood out. The amount of blood the heart pumps out per minute is called the cardiac output; it is a function of the number of heartbeats per minute and the amount of blood that the heart pumps out per beat.

Normally, the cardiac output is 5 to 5.5 liters of blood per minute and is related to overall oxygen demand. When you need more oxygen, as when you are fleeing the bad guys, the heart rate can increase up to threefold, and by constricting various blood vessels, the amount of blood in circulation increases twofold. More oxygenated blood in circulation allows for increased activity in the cells, which can then function more effectively to allow you to escape.

But in a nonemergency, as when Annie is bottom up, there is no need for increased oxygen. Dr. Horacio Kaufmann, a prominent New York University researcher specializing in diseases of the autonomic nervous system, explains how the bottom-up approach alters heart rate. As the volume of blood in circulation increases, Annie's heart rate automatically slows down so that the cardiac output remains constant. Her heart now beats slower, as it pumps out more blood per beat than when Annie is standing upright or even when she is lying down. Her breathing rate also automatically slows, as a larger volume of blood floods her lungs with each inhalation, allowing for more efficient oxygenation.

The net effect of Annie's bottom-up posture is that, without any direction from her frontal lobes, her heart rate and her breathing slow down. Deep, slow breaths and a slowed heartbeat send a strong message to her core brain, signaling that there is no threat in the environment, and the world is a safe place. The vagus nerve conveys the state of her body to Annie's core brain, which notifies her frontal lobes that Annie is calmer. And just like that, Annie *is* calmer.

Because her blood flow increased, Annie's heart needed to beat less rapidly to provide nutrients to her brain. The automatic reduction in heart rate means that her brain was tricked into thinking that Annie was relaxed. This bottom-up feedback from body to brain—which eases Annie into a state of calm—operates constantly in all of us even though we are rarely consciously aware of it. The state of flux, the constant tug of war between the relaxing influence of the parasympathetic vagal system and the alerting effect of the sympathetic adrenal system, allows us to respond appropriately and speedily, not only to the demands of our body but also to danger and reward in the environment.

Even though you don't register the bottom-up feedback from body to brain, there are ways to become aware of this parasympathetic control of your body. After many years of focused practice, yogis and experienced meditators can slow their heart rate and breathing by consciously asking this of their body. In other words, with experience, the rational frontal lobes can persuade the parasympathetic system and the core brain to "relax." However, most of us do not have the luxury of time or the dedicated discipline to spend hours training our brain to relax by practicing yoga or meditation.

Topsy-Turvy Treatment

My patient Lisa was getting more and more dependent on drugs, which were increasingly ineffective, and she still suffered from migraines and runaway anxiety. People afflicted with this type of headache often have neck and back problems, which promote a vicious cycle of chronic pain. Lisa was a typical example of this cycle, and I was worried about her and her crippling complaints.

Would Lisa respond to the bottom-up approach to calm? What if I literally forced her to hang upside down? Would she become calmer despite herself? Would her core brain, her parasympathetic system, and her vagus team up to override her frontal lobes and her sympathetic hyperactivity? Providing a type of reverse traction could certainly help, and it wouldn't make her headaches or back pain any worse.

We set about ordering an inversion table for Lisa. This piece of equipment, widely used by physical therapists to treat back ailments, offers a simple way to safely tilt and hang upside down. I must admit that both Lisa and I were skeptical that this would solve her problems, but we agreed it was worth a try. And I was eager to see if my hypothesis about "bottom-up therapy" would turn out to be true—at least in Lisa's case.

Lisa returned to my office three weeks later. "The inversion table is working," she announced with a huge smile. "Whenever I was stressed, I used the table, and now my headaches are much better. I lie on it for about five minutes; it's not like ta-da, everything's perfect, but I feel much better. When I sense a headache coming on, I get on the table and"—she tilted her head and shoulders back—"aahhh . . . it feels so good!"

Over the next three months, Lisa made more and more use of the table. To the astonishment of both of us, she was able not only to rid herself of her headaches but to stop using painkillers and Prozac.

And there was another intriguing aspect to this story. For several years, Lisa had been taking medication to reduce irritable bowel symptoms. After starting her bottom-up strategy, she noticed that her gastric problems lessened, and soon she was able to come off this drug as well. The last time I saw Lisa, she proclaimed, "I feel better than I have in years! And I am not popping any pills! Who would have believed it?"

What is it about an inversion table that promotes this type of transformation? Why does it work? I believe it has to do with satisfying core-brain needs and turning down the strident "frontal lobe on steroids" circuit. The good news is that all our systems have our well-being at heart. The bad news is that, if you are not aware of the brain's powers, you could, much like a country ravaged by an internal war, fall prey and succumb to illness, a sense of unease, a pervasive anxious feeling, and a lack of calm. Lisa was in that sorry state until she tried the inversion table.

Lisa's results were so dramatic and the action required to achieve them was so simple that I was reminded of Ellen, my dog trainer, and how she calmed Wolfie, my automobile acrobat. Does a calm body equal a calm mind in humans as much as in dogs? This may seem a reductionist question, but the truth is that, if the core brain is similar across mammalian species, as we know it to be, then it makes plenty of sense.

Vagal Tone

Standing on your head is, of course, not the only way to engage your parasympathetic system, increase your vagal activity, and thus embrace calm. If it were, we'd all be turning ourselves upside down every chance we got. In fact, people who are calm much or most of the time, or are able to calm themselves easily, have higher baseline vagal activity to begin with. The higher the vagal activity, the slower the heart rate and the better coordinated your heart rate is with your respiration. When you breathe in, your heart rate quickens to allow more efficient exchange of incoming oxygen and removal of carbon dioxide, and when you breathe out, your heart rate slows to conserve the energy expended on your pumping heart.

People with higher vagal activity are better at keeping their heart rate in tune with their inhalations and exhalations, allowing for efficient and calm breathing. This synchronization between heart rate and respiration is called respiratory sinus arrhythmia, or RSA, and is used as an index of overall vagal activity or vagal tone. One of the reasons that yoga can have a salutary effect is that it focuses on the calming process of deep rhythmic inhalations and exhalations.

Another way to evaluate vagal activity is to assess heart rate variability, or HRV, that is, the change in heart rate from beat to beat. The heart receives input through the parasympathetic vagus, which slows the heart down, and the adrenergic sympathetic system, which speeds it up. The ability to modulate heart rate at such a fine scale allows you to respond rapidly to the environment—for example, your heartbeat quickens when you see your beloved walk into the restaurant, even as you remain seated at your table. People with high HRV have a better balance between the two systems than those with low HRV, who have a sympathetic dominance. This dominance predisposes them not only to anxiety, panic, and depression but also to life-threatening heartbeat irregularities, heart attacks, and even death.

Is it possible to know what your basal vagal activity is? Measuring your vagal tone is not a simple task like checking your pulse. It is not expressed as a number; rather, it is a measure of how *responsive* your vagus is to changes in the environment and how quickly it can restore your body and brain to a resting, calm state after a perturbation. Vagal tone is assessed in research settings by recording the electrocardiogram and then measuring the variation in heartbeat with breathing, as well as the changes in the heartbeat from one beat to the next. While a steady pulse in an anxiety-provoking situation is a good indication of overall calm, it is not a good measure of overall vagal tone. However, the concept of slowing the pulse so that it tricks the brain into thinking that all is calm is exploited in the use of heartbeat-slowing medications like propranolol, which is sometimes used to treat stage fright.

Vagal tone is not simply an individual concern. In fact, all cultures aspire to a state of high vagal activity, notes Dr. Kaufmann, the New York University neuroscientist. Meditation in any form—with rosary beads, through yoga, by reciting incantations, or practicing traditional Buddhist techniques—ultimately achieves calm by increasing vagal activity. Increased vagal activity turns down the alerting adrenaline-driven sympathetic system and the frontal lobes, creating an internal state of calm. When researchers measured vagal activity across different cultures as subjects recited the rosary, meditated, or chanted, they came up with the same ideal number: Maximum vagal activity occurs when you take six breaths a minute. At that measure, the brain is most "at peace."

Is reducing your number of breaths per minute the way to raise your vagal activity? The answer is not clear. But we do know that accomplishing this is likely to be daunting. Attaining this serene state requires many years of practice, as the usual number of breaths is twelve to fifteen per minute. Many meditative traditions use focused and slow breathing to attain calm. But as anyone who has tried and failed at this will confirm, attempting to slow your breathing while your frontal lobes are chattering on—about "shoes and ships and sealing-wax, of cabbages and kings. And why the sea is boiling hot, and whether pigs have wings," as Lewis Carroll inimitably put it—can be frustrating. Optimal breathing with this type of conscious meditation is likely to be foiled by your busy frontal lobes, which seduce you into thoughts far removed from inhalations and exhalations.

Nonetheless, it is widely known that deep breathing and meditation promote relaxation and a sense of inner calm. What is less known is that these techniques accomplish this by ramping up the activity of the vagus nerve. In a state of high vagal activity, the vagus nerve is more stimulated, resulting in slower breaths, slower heart rate, reduced bowel irritability, and better synchronization of heart rate with respiration. This results in an optimal body state. Through bottom-up feedback, the core brain is calmed, which relaxes the frontal lobes as well.

An abundance of research supports these observations. In one study, sixty-three Romanian university students between the ages of nineteen and twenty-four were given a questionnaire to determine their baseline anxiety in response to threats. Thirty-six of the students scored high, reflecting a higher level of baseline anxiety; the remainder scored low. All students were then given a difficult time-limited arithmetic task, found to be a universally anxiety-provoking exercise. After the test, they engaged in breathing and meditative exercises and returned to a state of "vigilant calm." Their heart rate and skin conductance (another way of assessing anxiety) were measured.

Those students who had high threat anxiety to start with—that is, those with higher baseline anxiety—were more likely to have reduced vagal activity. This showed up as lower heart rate variability (HRV) after the stressful mental task. Interestingly, the meditative and breathing exercises increased the HRV for both anxious and mellow students. Higher threat anxiety was associated with reduced HRV, which is a reflection of the reduced responsiveness of the vagus to the environment.

The optimal state of high vagal activity not only helps us attain a state of calm; it also allows us to be better members of our community, because a brain and a body in this relaxed state is better able to interact with others. Good vagal tone helps us adapt well to our environment, so our body is able to amp up in the face of a threat but is just as able to calm down when the "all clear" sounds. My grandfather provides a practical example of the value of high vagal tone. The vagal "brake" on his sympathetic nervous system functioned so efficiently that when he was faced with a granddaughter's ire-provoking shenanigans, he could manage the situation calmly—and move on to dealing with less asinine matters!

The significance of community and its relevance to calm is borne out by the importance the brain places on dealing with other people, based on the multiple brain areas allocated—in some cases almost exclusively—to this task. Earlier, I described the core brain's hardwired circuits and pattern generators, which let us communicate without rational and conscious thought. Additionally, "mirror neurons" within the evolutionarily newer parts of the brain are activated in patterns similar to what we observe in our fellow human beings. When you watch a friend eat ice cream or cry or laugh politely at an unfunny joke, mirror neurons in your brain allow you to mimic the experience both in intent and in action. A part of your brain relishes the ice cream, cries, or laughs politely at the unfunny joke. "Laugh and the world laughs with you; cry and you cry alone," the old adage tells us. Why does crying isolate you instead of drawing you closer to others? One reason is that mirror neurons and the core brain make others experience your sadness and thus avoid you.

Part of our brain is devoted exclusively to the recognition of the faces of other humans, which speaks to how important socializing is for survival. In an interesting study of a physician whose entire higher visual brain was destroyed by a stroke, his core brain could distinguish between angry and sad faces even though it could not tell the difference between a circle and a square. There are also parts of the brain devoted solely to understanding changes in the inflections of our words, distinct from the meaning of words themselves. To our brains, no man is an island.

The vagus nerve fibers that influence breathing and heart rate are very close to those nerves that influence facial expression and the production of sound and speech, providing further evidence of the connection between human communication and body functions at a core-brain level. For example, studies have shown that lower vagal activity is correlated with avoidant behaviors, anxiety, and depression in children with no mental disorder, as well as in those with clinical anxiety and depression. In one study of 160 third graders, when children identified as being in a state of "anxious solitude" were exposed to social rejection, they were more likely to experience low vagal activity and sustained increases in heart rate. This pattern is also seen in clinically depressed and anxious adults.

The more fit our cardiovascular and respiratory systems are, the better able we are to mold and train our vagus to synchronize and adapt to the environment, giving us higher vagal tone. All of us probably had higher vagal and parasympathetic relaxing abilities as children, which diminish as we grow up. This is true even though, as many a parent can attest, some babies are born with low vagal activity (cranky, hard-to-soothe infants) as compared to "happy" babies who, like Walt, are easy to comfort.

The extent to which we reduce our vagal activity and our ability to relax as we grow older is shaped by the nature of our infancy and childhood. Simple things like aerobic exercise promote higher vagal tone in children, helping them be more calm and relaxed. And this ability tends to persist into adolescence and adulthood. Infants and children with high vagal tone are better able to focus; those with low vagal activity are more prone to stress and less able to react to the environment. In one study, in which children were followed for six years, from age one to seven, researchers found that those with high vagal activity as babies were also more adventurous by twenty-one months; more talkative, smiling, and sociable by age four; and continued to smile more and be more spontaneous in conversation at age seven. Infants and young children with high vagal activity are more responsive to the environment and have superior ability to shift and sustain attention. This is no surprise. When you are relaxed, it is easier to pay attention to the world around you than when you are afraid and anxious.

Children and animals use different techniques—thumb sucking, belly rubbing, nuzzling, stroking, and just playing—to soothe themselves. Chimpanzees spend hours on calm-inducing grooming activity, primarily in the form of nit-picking, and there is evidence that villagers in the Middle Ages engaged in this activity as well, for the same reason. Human babies up to three months of age need their parents to put them to sleep or to comfort them when they wake up during the night. But by the age of one year, they have learned to soothe themselves well enough to get back to sleep on their own. Rarely do babies sleep through the night, generally waking up a few times. Sometimes, when this happens, babies cry, sometimes they coo or babble to themselves and then drift back to sleep. Although no one knows why this is so, girl babies are more successful than boy babies in soothing themselves when they awaken. While self-soothing babies wake up just as often as babies who need parental attention, they are more likely to just lie quietly than to vocalize.

Many of these methods of self-soothing are no longer available to us as adults. Yet, touching and social interaction are two common and very effective ways for us to continue to calm ourselves. Human contact in empathic, trusting situations releases hormones like oxytocin that allow people to bond with each other, thereby satisfying the core brain and increasing one's sense of calm. Physical closeness reduces levels of cortisol, a stress hormone, and adrenaline, a hormone that activates the sympathetic system, thus promoting a sense of well-being and relaxation.

Vagal activity is also related to malfunctions of the gut, which commonly undermine a sense of calm. We saw how Lisa's irritable bowel was relieved by her sessions on the inversion table. When a stomach upset leaves an infant listless and apathetic, or an adult irritable, the vagus/core-brain circuit is responsible. Information from the gut is conveyed to the core brain, which then responds by altering the activity of the intestinal system. With anxiety, our vagal brake malfunctions and we are unable to tamp down our irritable bowels.

How is the bottom-up interplay between vagal feedback to the core brain and from there to consciousness and back down through the feedback loop altered in anxiety? We know that anxiety exacerbates bowel mobility by altering sympathetic and parasympathetic input to the gut. My friend Paul suffers severe diarrhea before every major sales meeting. A few hours before these meetings, he starts running to the bathroom, the pace quickening until he has to go several times in the hour before the meeting begins. Paul's anxiety produces bowel irritability, which in turn feeds back to the brain a sense of anxiety, creating a sickening cycle that is reinforced with each monthly sales meeting.

We also know that in rodents, damage to the vagus nerve lessens or prevents specific sickness behaviors by removing the vagal brake and reducing vagal tone. This causes changes in motivation and social behavior. Once again, the bottom-up dynamic is in play. Information from the body is conveyed up the spinal cord and integrated with unconscious information received from the vagus, finally traveling to a place in the higher brain, called the insula, to register as illness or wellness. This may be the crucial place where we consciously perceive how we are and how we feel about how we are. This is where the real answer to the question, "How *are* you?" resides. When we answer rationally, we are using the frontal lobes to convey information, so we are able to smile and say, "Fine, thank you," even as we feel like death warmed over. Children and animals, on the other hand, cannot do this. With less well-developed frontal-lobe rationalization processes, they are more in tune with their guts and thus behave the way they feel: well or unwell.

A fascinating procedure called vagal-nerve stimulation has been shown to relieve depression, sleep difficulties, and even obesity. In this process, electrodes are attached to the vagus nerve as it passes close to the skin near the neck, causing the electrical activity from the vagus to be "upregulated," or bolstered. This method "informs" the core brain and the higher brain areas, including the frontal lobes, that vagal activity is "high"—thus mimicking a state of calm.

Vagal nerve stimulation can help obese people eat less by creating a sense of satiety. It can also reduce seizures and anxiety in those who suffer from epilepsy. There is a good deal of evidence that this method works. Functional imaging studies of people undergoing vagal nerve stimulation show changes in brain activity all the way up to the level of the frontal lobes.

The vagus also plays a vital part in conserving energy. The importance of vagal stimulation in managing our high-energy-using respiratory and cardiovascular systems, as well as the crucial role the vagus plays in toning down the adrenergic sympathetic system, speak to this vagal function. Optimal energy management is important to our bodies. It is achieved by physical and mental enrichment through the vagal-parasympathetic system, and physical and mental energy expenditure by the adrenergic sympathetic system. These systems are integrated and managed at an unconscious level by the core brain. This optimal energy management was what came across to me as my grandfather's equanimity—the calmness of mind and body that I longed for.

In nature, a "calm and social connection system" opposes the "arousal stressed" system, which are both mediated through the core brain. Positive social engagement activates the parasympathetic vagal system and can suppress the sympathetic adrenal system's arousal response, thus reducing stress and anxiety, promoting calm and feelings of belonging. Increasing vagal activity improves sociability, immune function, and positive emotions, and it suppresses pain. Having read all of this, don't you want to put this book down and stand on your head?

Could It Save Your Life?

Could increased vagal activity have saved the life of Kenneth Lay? Lay was CEO of the now defunct Enron Corporation, earning over forty million dollars a year. In May 2006, when he was sixty-four, Lay was indicted on charges of fraud; sentencing was scheduled for September 2006. His life in shambles, Lay was vacationing in Colorado that July when he suffered a massive heart attack and died. Was Kenneth Lay literally scared to death? Did anxiety kill him?

Martin Samuels, professor of neurology at Harvard Medical School, studies sudden death, also known as "voodoo death." He suspects fear and anxiety may have had something to do with Lay's demise. When someone dies suddenly upon hearing bad news, or when their heart stops beating because they mistook a shadow for an intruder, they may be victims of voodoo death. Samuels finds that a massive surge in the sympathetic system's adrenaline kills cells in the heart, causing it to stop seemingly instantaneously. Voodoo death is similar to the "broken heart syndrome," in which high levels of adrenaline are also found. In this syndrome, women have symptoms of chest pain after a stressful event such as a breakup, and tests show the kind of damage seen in heart attacks. Ninety percent of the sufferers are women and nearly all recover completely.

Fear, panic, extreme anxiety, and terror all activate the adrenergic sympathetic system and increase heart rate. However, this view of anxiety overlooks parasympathetic involvement. Higher vagal-parasympathetic activity can counteract sympathetic excitability, protect against a runaway heart rate, and stabilize a sympathetically charged overactive system. Extreme anxiety can cause sudden death by its effects on the heart through the actions of adrenaline and the sympathetic nervous system. But high vagal activity could prevent this from occurring.

Did Kenneth Lay have low vagal activity and reduced parasympathetic activity, which prevented him from reining in a runaway sympathetic system? Do women who suffer from broken heart syndrome have low counterbalancing vagal-parasympathetic activity? We don't know the answers to these questions, but we do know that people who suffer from panic attacks and those who faint at the sight of blood have lower vagal activity than those who don't. The point is that the *balance* of the alerting sympathetic system and the calming parasympathetic system, with integration at the core brain, is more important than any one component. To survive, it is important to react to the environment with an appropriate amount of stress or calm. Both states are crucial in helping us navigate the world.

This interplay is seen in the extreme as well. When the parasympathetic system is dialed all the way up and the sympathetic all the way down, we go into a *savasana* state, which, translated from Sanskrit, literally means the "death pose" in yoga. In this phase, metabolism slows down and the body assumes a stillness, a limpness, a prelude to the state of death. In states of extreme threat—when fright, flight, and fight are no longer possible—the "roll over and play dead" stance can help save our lives.

Stephen Porges, a professor of psychiatry at the University of Illinois, goes a step further in integrating the vagus, core brain, social engagement, and calm. He observes that humans must differentiate friend from foe to determine if an environment is safe. We need to decide quickly whom to trust and whom to run from. We need to do this "instinctively," well before our rational frontal lobes are activated. If we don't run quickly enough, the spear or the bullet will get to us faster. Porges postulates a social engagement system mediated by the vagus, which feeds directly back into the emotion-producing facial and voice muscles and that allows us to respond emotionally and automatically, without interference from our frontal lobes.

When we realize and integrate at the core-brain level that the person confronting us is the enemy, our vagal tone decreases, our face contorts, we scream, our heart rate increases because the sympathetic influence is unopposed. We breathe faster and we are running before we know it.

We saw this in the story of the woman who fled in high heels from would-be attackers and didn't notice her broken heel until she had hobbled home to safety. There her parasympathetic system kicked in and returned her to a state of calm. Her heart rate slowed, she breathed more slowly, and she broke into a smile—again, before she "knew" it.

As we've seen, our core-brain/vagal circuit is wired not just to keep us safe from the elements but also to protect us from the wrath of our fellow men. Not just to keep us healthy but also to maintain connections to our community. In fact, our brain responds to the threat of losing human companionship in the same way it does to the threat of famine, with stress hormones flooding the body, creating anxiety, and destroying calm.

If the core-brain/vagal circuit is so important, why did evolution give us stress? Why were we not born with only the calm circuits so we never have to deal with anxiety and agitation? Why can't those responses be modified so there is just the right amount of stress? Indeed, why do we need stress at all? The surprising answers to these questions mean an awful lot to the past and future of our species—which includes your grandchildren.

CHAPTER 4

Why Ever Stress?

The year was 1844. The explorer and medical missionary David Livingstone, thirty years old, was living in Mabotsa village, an area teeming with lions, in what is now South Africa. One day, a lion killed nine sheep in broad daylight on a hill just opposite Livingstone's house. The villagers ran out to try to trap the lion, and Livingstone tells us, "contrary to my custom, I imprudently went with them," cheering them on as they tried to slay the lion. Although they surrounded the beast and managed to wound it, the poor animal broke through the human circle and escaped.

Livingstone headed back home after all this excitement, along with the villagers. He recounts what happened next in a letter to his father: "I heard a shout starting, and looking half round, I saw the lion just in the act of springing upon me. . . . The lion caught me by the shoulder and we both came to the ground together. Growling horribly, he shook me as a terrier dog does a rat. The shock produced a stupor similar to that which seems to be felt by a mouse after the first shake of a cat. It caused a sort of dreaminess in which there was no sense of pain nor feeling of terror, though [I was] quite conscious of all that was happening. It was like what patients . . . under the influence of chloroform describe, who see all the operation, but feel not the knife. . . . The shake annihilated fear and allowed no sense of horror in looking round at the beast."

Reflecting on his experience, young Livingstone went on to observe: "This peculiar state is probably produced in all animals killed by the carnivore; and if so, is a merciful provision by our benevolent creator for lessening the pain of death."

Diverted by the natives, the lion eventually died from its wounds, and Livingstone lived to tell the tale. His left arm was crushed into splinters and healed poorly. In fact, the marks from the attack were used to identify his body when he died decades later, having lost contact with the outside world. To someone who asked earnestly what his thoughts were when the lion was above him, Livingstone replied, "I was thinking what part of me he would eat first"—a grotesque thought, which some people considered strange in so "good a man," but which was quite in accordance with human experience in similar circumstances.

Livingstone's story is not only a dramatic example of the

bottom-up mechanism of calm but also a vivid demonstration of both the stress response and the reason for its existence. The stress response flourishes because it keeps us alive as it kept Livingstone alive. In the beginning, as he became aware of the lion, there was discharge of his sympathetic system, and the fright and flight aspects of the stress response were deployed. However, when confronted by the overwhelming nature of the situation, with the odds against him, the parasympathetic part of the stress response took over and Livingstone became completely calm. His body went limp as his vagal tone increased. His heart rate likely slowed, his breathing likely became shallow. His core brain, which shares so much in common with other mammals, produced in him "a stupor." In this relaxed state, his efficient stress response presented Livingston as a less tantalizing, more inert prey to the lion, allowing for the precious extra time required for the Mabotsans to rescue him.

The Origin of Stress

We cannot address whether or not the stress response is a "merciful provision" of a "benevolent creator," as Livingstone wrote, but we do know that it is a product of evolution, designed to keep us alive. Over millennia and across species, evolution has conserved our body's stress response because it is what allowed our ancestors to survive and thus to pass on their genes. You are reading this book today because your ancestors were better than their peers at generating the appropriate stress responses necessary to survive infections and predators. Your forebearers pumped out the right amount of the stress hormone cortisol at the right times, adjusted their sympathetic and parasympathetic systems in just the right way to obtain enough food to survive, to have enough fear to flee at the right time, and enough courage and wit to come up with the right prehistoric one-liner at the prehistoric singles bar to charm the right person into procreating.

Despite its crucial role in preserving the organism and the species, the stress response is always a trade-off of risks and benefits. Imagine, for example, a herd of prehistoric giraffes drinking at a watering hole. The giraffes that are particularly wary of predators will speed from the hole at the sight of a lion. They won't die from a lion's attack, but they may perish from dehydration, having never been able to slake their thirst. Giraffes that have not learned to fear lions will never die of thirst but are likely to be a lion's entrée for the day. The giraffes that survive and go on to breed have the right amount of the stress response, just the right balance between fear of lions and need for water.

Scientists are constantly trying to one-up evolution by tinkering with the stress response. Stress resistance, whether in the form of drought-resistant crops or fatigue-resistant athletes, is a prized attribute. But science has taught us that tampering with our innate ability to cope with threats by modifying a stress response may well involve a sacrifice in function in another, often unexpected area. And this sacrifice, or trade-off, can occur generations after the modification is introduced.

For example, consider some interesting investigations of life extension that have been conducted using Drosophila melanogaster, also known as the fruit fly. Is it possible, the scientists asked, to introduce stress into the insects' lives and thus produce hardier and longer-lived flies? And, if so, what would this tell us about the stress response and human longevity?

Researchers like to work with these tiny flies because their genetics is known, they can be grown and manipulated easily in a lab, and multiple generations of fruit flies can be produced in a few months. (Also, PETA demonstrators rarely turn up outside laboratories, demanding the release of these insects.) Drosophila can be bred to withstand the stresses both of starvation and of thirst ("desiccation," in scientific lingo). These superflies, which are three to six times more resistant to starvation and desiccation, are raised in test tubes with reduced or no access to water or food. Fast-forward sixty Drosophila generations, approximately sixty months. Although the offspring of the superflies were longer living, they were also less fertile and their larvae less healthy than the larvae of their more stress-susceptible, starvation- and desiccation-prone cousins. In other words, despite producing longer-lived, more individually hardy fruit flies, scientists sacrificed both the fertility and the viability of the flies' offspring, making them less likely to survive into the future.

Even though they are "hardier" in the immediate and even the next generations, over time this line of fruit flies will die out. (On the other hand, dying out may not necessarily be a bad stress response from the fruit fly perspective if you are going to spend your days flying around in a test tube without food or water.) Clearly, modifying a stress response to produce a benefit—in this case, longer life—can backfire.

It is impossible to perform such studies in human beings, for obvious reasons. However, there have been natural conditions in which scientists have been able to observe children whose parents have survived periods of famine. In the remote northern Swedish town of Överkalix, where the most recent census still records a population of less than a thousand, there were severe food shortages interspersed with normal food supply in the early 1800s. The 1,626 direct descendants of the 271 children who were between the ages of nine and twelve during those years were examined two generations later, to understand the possibly heritable effects of the stress response to starvation. Amazingly, not only were the grandchildren of those exposed to famine living as much as a decade longer than the grandchildren of those who had not faced starvation, they were also less likely to suffer from heart disease. This speaks to either transgenerational karma or an alteration in the expression of genes coding for longevity and heart disease in the face of stress among those exposed to famine during crucial childhood years. These traits were then passed down two generations.

It is important to realize that the stress response allows us to survive not just physical stressors but also mental stressors. We all know of a physical stressor causing an emotional stress response—darkness making one afraid, for example. Physical stress can predispose children to mental illness even when they are in the womb. Women who were pregnant during the severe famine that accompanied the Nazi blockade of Holland at the end of World War II were more likely to give birth to babies who later developed schizophrenia.

The stress response affects the survival of the individual and, more important, the survival of the species. When we tamper with the stress response, payoffs that initially seem positive, such as individual longevity, may ultimately harm the species by affecting fertility, as in the case of our unfortunate fruit flies. The complete effects of the modified stress response in human beings, as in the descendants of the Överkalix and the Dutch famine survivors, may need several generations to become apparent.

How does the stress response help us survive as human beings in situations other than lion attacks? To answer this question, we must look at the evolution of stress over different species, including humans. All animals with a backbone appear to produce stress-type steroid hormones. Even reptiles, insects, and some worms make a substance nearly identical to steroids. This substance is not only closely linked with other stress-related chemicals like adrenaline but also with infection-fighting chemicals. That these hormones and brain chemicals have remained unchanged over hundreds of millions of years in a multitude of life-forms speaks to their tremendous importance in keeping us alive.

Steroid hormones are released—and the energizing, sympathetic system activated—not only in response to arousal due to threat, as was the case with Livingstone, but also from arousal due to opportunity (food or sex, for example). And the classic stress hormone, cortisol, is also secreted in response to rewards and pleasure. Livingstone's cortisol level was high because he was attacked; the lion's was high because he found prey. The important point here is that whether you are thrilled to be making love or are eagerly anticipating a terrific meal, whether you are nervous about delivering a speech or terrified that you'll be eaten by a wild animal, your body responds in the same way: pumping up the adrenaline, pouring out cortisol, getting your heart to beat faster, and leaving you breathless with excitement.

The sympathetic nervous system, which is responsible for the release of adrenaline and cortisol, is counterbalanced, as is everything in our bodies, by the equal and opposite parasympathetic nervous system, aided by the vagus nerve. When Livingstone was in the jaws of the lion, and all hope seemed lost, his "fight or flight" sympathetic system was deactivated and his "rest and repair" parasympathetic system came into play. It made no sense anymore to fight (how could he?), there was no option for flight (how could he?), and so Livingstone's parasympathetic system and vagus nerve took over. He became limp and lost all fear. In the face of death under horrifying circumstances, Livingstone's vagus nerve and parasympathetic system suffused him with calm.

David Livingstone's stress overtook him suddenly and unpredictably. And his calm was engendered more by nature—mediated by his core brain—than by nurture. But nurture can also contribute to our vulnerability to stress. Consider the macaque, one of our closest monkey cousins. Researchers studying macaque infants found that when they were raised with mothers in environments of either constant want or constant plenty, they were less anxious than when raised in conditions in which periods of feasting were randomly interspersed with famine. In other words, it was not so much the poverty of the environment that determined the infants' anxiety and stress but the inability of the macaque mothers to respond predictably to the needs of their babies in these uncertain situations. Our core brains, like those of the macaques, crave consistency.

Trying to link a behavior like a macaque baby's level of anxiety, which is the end result of a multitude of factors, to a simpler measure, such as cortisol levels, is always risky, but it helps shed light on the complexity of the stress response. Early life experience seems to modify adult stress response and the release of cortisol. Adult rats that had more attentive maternal care during the first week of life had a more sensitive chemical "switch" that turned off cortisol production once a sufficient amount had been released into the system. Rats that had less maternal care as pups had stickier switches, requiring a lot more cortisol in their system before they turned down production.

Both the macaque and rat experiments point to what Charles Darwin meant when he said, "It is not the strongest of the species that survives, nor the most intelligent that survives. It is the one that is the most adaptable to change." The core brain needs to be appropriately adaptive to stress, as poor adaptability floods the body with constantly high and harmful levels of cortisol.

While the stress response is essential to survival, it is also expensive, both calorically, as the body prepares itself for a possible attack, and in its effects on the body, even at normal doses. Being in a constant state of vigilance is not conducive to feeding and mating, which are crucial for survival of the species. Yet, as we have seen, not having a stress response is even more expensive and can cost one's life. The trick is to distinguish the inconsequential from the true threat. While my grandfather may have responded with a higher level of cortisol to the annoying lack of drinking water caused by his granddaughter's prank, his baseline level of calm quickly enabled him to turn down the stress response. Compare my grandfather to my patient George, who suffered debilitating long-term problems because he couldn't control his runaway reaction to stress.

The Daily Flow

George is a forty-year-old man sent to me by his internist for numerous complaints, some of them vaguely neurological. "I am not sure if you can help him," said his internist, Sharon, when she called to refer him. "I'm not sure anyone can."

When I met George, he was quick to describe his problem. "I have 24/7 aches—in my arms and legs—for the last eleven years," he said. "My symptoms get worse, usually more with mental stress than physical stress. I start to ache all over. I have trouble moving my bowels. I feel nauseous and have trouble thinking or coming up with the right words. If I concentrate for more than a couple of hours, I can bring on my symptoms."

George was sure these problems were work related. When he was in his twenties, he had worked on a trading floor, one of the highest of high-pressure jobs. It was there that he first developed similar symptoms. This had prompted him to quit his job and go to business school. He was healthy for the next several years, but when he once again took a high-pressure trading-floor job, his symptoms returned. This time around, though, a curious thing happened. His symptoms "developed a mind of their own." They continued plaguing him after he eventually quit working. In the four years since stopping work, he said, "I have an overwhelming ill feeling. I feel like I'm being poisoned."

George has been given various diagnoses, including autoimmune disease, chronic fatigue syndrome, and chronic Lyme disease. He had been on high doses of immunosuppressive agents for months at a time and on antibiotics for a whole year, in what turned out to be vain attempts to treat his symptoms. He had, to no avail, tried nine different antidepressant medications and a series of Valium-type medications.

It was my opinion that George could no longer turn off his stress response switch efficiently. I did not measure his cortisol level or his ability to adapt his stress response to different environmental threats, nor did I measure his vagal tone, since my office is not equipped to do this. But I am confident that had I been able to do so, I would have found an imbalance in George's stress response, with a hyperactive sympathetic system and a less active parasympathetic system and reduced vagal tone.

George was able to turn off his response the first time he was exposed to chronic, overwhelming stress, but the second time, he was less resilient. His switch became refractory and malfunctioned; he became, to put it in Darwinspeak, less "adaptable." This type of chronic response to a real or imagined *mental* stressor, which so many of us experience nowadays, is very different from the acute, real or imagined, *physical* stressor that our core brain is so well equipped to deal with.

Livingstone's response to the lion's attack, his "stupor" and "dreaminess," is another reason stress mechanisms survive in all of us and why they are so important. The stress response alleviates pain. People who have endured terrible suffering often say they felt "numb," a state that allows someone to function after tremendous trauma. Although it was not a conscious decision, and was mediated entirely by his core brain, Livingstone protected himself by going limp as the lion shook him. If instead he had tried to wrestle the lion or attempted to flee, he might have been fatally injured. Moreover, the lion's shake "annihilated fear," his stress response allowing Livingstone to essentially self-anesthetize, "like what patients . . . under the influence of chloroform describe."

The value of the stress response in reducing physical or psychological pain has been documented in the laboratory as well as in real life. In rats that are painfully stimulated without the possibility of escape, the response offers the rodents relief by reducing their sensitivity to all types of pain, not just the pain used in the experiment. This is not as far-fetched as it seems. Think back to a time when you or someone you know was "numb" from the pain of the loss of a loved one, or from a romance gone bad. Such a person is less sensitive in other ways as well: They may not get a joke; they may react less to other hardships; they appear "hardened."

The same stress mechanisms operate in your body whether you are being mauled by a lion, coping with losing your job, mourning the death of your best friend, or celebrating your birthday. This inability to fine-tune the body's identical response to varying stimuli is what gets us into trouble, as in George's case. When your stress response is working well—for example, when, after George's first stint at his job, he quit to go to business school and recovered fully—you go on with life without suffering ill health. When it goes awry, then symptoms of disease develop.

George's story makes clear that the right amount of stress response is essential to keep us healthy. And that's another reason that the stress machinery has survived through evolution and across species. Appropriate release of stress hormones helps us fight off disease conditions that threaten to destroy the body. Today, when modern medicine is confronted with overwhelming infection or inflammation, physicians resort to an age-old cure devised by nature itself—they place the patient on high doses of steroids.

Despite its value as a defense against pain and in keeping us healthy, stress does not get good press. Lisa's "stress-filled" life has given her migraines and turned her into a "complete mess." Magazines and Web sites are filled with "cures" for stress, promising to keep you calm even under the most harrowing circumstances. Stress is looked upon unfavorably because it accompanies distress or suffering, even though the stress defense itself is essential to combating that suffering. But, as the pioneering evolutionary scientists George Williams and Randolph Nesse have pointed out, the very feeling of misery associated with stress and illness is adaptive.

Vomiting and coughing when you've swallowed something you shouldn't have may save your life by expelling the harmful pathogens. The flu leaves you hot and achy, with a runny nose and miserable cough, but at the same time, it allows your body's defense mechanisms to purge the invading virus from your system and sends you to bed to recover in peace. Blocking a cough during a lung infection is not a good idea, as the cough clears your respiratory tracts. The mucus in your runny nose contains powerful antiviral agents, and its viscosity also protects your system from foreign germs.

Even as we complain about stress and yearn to eliminate it from our lives, we cannot help but recognize that stress in moderation is a marvelous thing. A little stage fright may lead to a more nuanced Act I, with more core-brain activation. And this will help you connect better with the core brains of the audience, bringing on more curtain calls. A little fear in a shady neighborhood arouses your senses, alerting you to possible danger.

Stress also sparks creativity and mental renewal, keeping you open to new concepts and ideas. It enhances performance, sharpens acuity, and improves memory. Athletes at the top of their game usually have increased levels of the stress hormone cortisol and some reduction in vagal tone. Studies have shown that being a little bit anxious as you study for an exam, when you'd rather be out with your friends, helps you remember what you learned and score better on the test.

Each year I deliver about forty talks to colleagues, students, and lay audiences. Over the years, I've discovered that I do my best when I'm slightly on edge, when I'm just a little worried or uncomfortable, especially when I'm presenting new material. When I'm too relaxed, I don't connect as well to the audience, I'm not as alert to cues from my listeners, not as responsive and perceptive. A tiny touch of anxiety helps me be nimble enough to change tracks if the situation demands it.

Delivering my talks in the core-brain space between fright and calm resonates with the psychologist Mihaly Csikszentmihalyi's concept of "flow." Based on his studies of hundreds of high performers, including Nobel laureates, he concludes that we perform best and experience the highest level of well-being and happiness when we are in a "flow" state of focused energy, striking a perfect balance between boredom and anxiety. From our perspective, this balance is between the parasympathetic and sympathetic, the right amount of vagal tone, the right balance between core-brain and frontal-lobe needs. Interestingly, Csikszentmihalyi's research makes short shift of the notion that work is necessarily drudgery and that leisure time is inevitably pleasurable. In fact, as I am sure we have all experienced, he finds that we are more likely to attain the "flow" state when we are deeply involved in tasks that require a high degree of skill and commitment, *regardless of whether we are working or playing.*

Although the term had not yet been invented, I believe my grandfather spent much of his rich long life in the flow state, as he energetically and calmly confronted challenges and brought a high level of expertise and dedication to everything he undertook. This optimal way of being always entails some measure of stress, and yet it epitomizes calm.

Eliminating the stress response is not possible, not desirable, and not compatible with life. So instead of "How can I stop being stressed?" we should be asking, "How can I reset my defense thermostat so that it doesn't switch into high gear when faced with a slight threat? How can I turn the stress response off quickly and efficiently when it is no longer needed?"

Stress Management in Adversity

The most important thing to keep in mind about getting to calm is that it is not about getting rid of stress; it is about *managing* it. Practices like yoga or meditation nourish the core brain and promote parasympathetic relaxation and increase vagal tone. Yoga's shoulder stand pose and the bottom-up position of the inversion table slow down the heartbeat and breathing rate, so that the core brain and the vagus can send an all-clear message to the frontal lobes—leading to calm. The *savasana*, or death pose in yoga, is a state of total relaxation similar to what Livingstone went into when he became limp. He was literally playing dead. But there are other, simpler ways of nourishing the core brain. Laughter is one way. Shared communion with a fellow human being. Touch. These and other means of relaxation help reset the stress thermostat and titrate your stress response.

Of course, there are always going to be small annoying consequences to the stress response, just as there are negative consequences from walking upright. Getting around on two feet left us with lower back pain, ruptured disks, and cesarean sections, but it also made it possible for us to salsa dance and jet off to the moon and try to walk on it. The stress response in normal amounts has some drawbacks, including discomfort, pain, fear, and sweaty underarm stains, but also vast advantages. Calm is understanding this balance and learning how to calibrate your stress switch, turn it on and to the right amount when you need it, and to turn it off when you don't. Calm is about learning how to *adapt* to our environment.

In the unlikely event that I am a "normal weight" model striding the catwalk for Givenchy, I am going to have to adapt my stress response every time someone says that I am fat or else I may become anorexic. If I am an editor in the increasingly high-pressure and high-stress world that is book publishing, dealing with difficult authors who renege on deadlines and bosses who are pushing to meet them, how do I reset my stress thermostat so that my cortisol levels are not constantly high? I do so by listening and responding to my core-brain needs and toning down my doomsday-prone frontal lobes.

Each of our brains also has a different baseline level of activity. Anxious people like Lisa or George engage their brain networks in a slightly different way from an upbeat person like David Livingstone. Our brains idle differently and therefore perform differently. We have the same computer but run individual versions of the operating system. Some of us are born calmer than others and some of us grow calmer than others, shaped by the vicissitudes of life, factors as diverse as our grandfathers' diets when they were nine or our mothers' care when we were week-old rat pups.

Contrary to popular belief, the brain is active all the time, even at complete rest. Scans that measure glucose use in the brain have revealed "default mode networks" that become engaged when the brain is not performing a task, essentially, when the brain is idling. These networks—which are detected in newborns as early as two weeks—are active when you are twiddling your thumbs, when you are daydreaming, or when you are just resting quietly. They help you integrate current needs with past experience and future goals. As we grow into adults, these networks help us mold and prune our baseline brain states, to become either more like George or more like my grandfather. Default mode networks, first described by Dr. Marcus Raichle at Washington University, may help define our baseline demeanors.

Stress disrupts the default mode networks. In the rats

that were subjected to pain, we can speculate that the default mode network became infinitesimally altered, perhaps preventing a baseline level of calm. Our experiences and environment may alter our default networks, activating slightly different background programs that allow us to react more or less calmly to the ups and downs of our lives.

Stress has been with us since the beginning of time, and while I have emphasized the stress of modern, urban life, the truth is that modern man has it made. Our ancestors competed for food while risking their lives every day. Infections were rife and without cure. Fractures and injuries were common. There was no protection against the elements, no laws against crime. While there was more social bonding in smaller communities, the organic threats of nature were surely devastating.

For most of us, life today is far different from that of our ancestors.

As evolutionist Dr. Randolph Nesse eloquently puts it, "Most stresses in modern life arise not from physical dangers or deficiencies, but from our tendency to commit ourselves to personal goals that are too many and too high. When our efforts to accomplish these goals are thwarted or when we cannot pursue all the goals at once and must give something up, *the stress reaction is expressed. In short, much stress arises, ultimately, not from a mismatch between our abilities and the environment's demands, but from a mismatch between what we desire and what we can have.*" [emphasis mine]

Annie, the "first-class worrier," is a perfect example. Forever fretting about money even though she's a top earner in her firm, constantly obsessing about unlikely catastrophes, Annie has condemned herself to living an angst-filled life. She would do well to surround herself with her loving family and have them share in her concerns and offer her comfort. By isolating her worries from them, her core brain communicates her anxiety to others but does not receive any solace in return.

Activating the core brain and the vagus is crucial in managing stress. A friend of mine, Tom, seemed to know this instinctively when he was fired from his position as a senior associate in a prominent law firm. He was given several weeks' notice and a decent severance package, but that didn't soften the shock or stop him from panicking. At first. Then, "almost magically," as he put it, Tom knew what he had to do. Attending to practical matters—like making sure his assistant would be taken care of—helped engage his core brain and build up his vagal tone by caring for another human being, in this case, his assistant. Then, instead of suffering alone or hiding his worries from his family, he drew even closer to his wife and children. Small things, ordinary intimacies like making dinner for his wife, reading to his young daughter, shooting hoops with his son, helped protect him from the negative effects of stress.

It is well known that people who lose their jobs or experience adverse events are more prone to illness and may withdraw from others, paralyzed by bitterness and anxiety.

But not Tom. He clung close to the people he loved and trusted and who loved him. What's more, he flooded his brain with happy chemicals and filled his stomach with good food. Tom drew on the sustenance of his "tribe."

Here's a true story of another man. Rising at eight each day, he would read for several hours in bed, scanning newspapers and poring over books. Morning meetings and dictation—often while he was still in his pajamas—would be followed by a long lunch and then a nap of an hour or more. He enjoyed leisurely lunches and sumptuous dinners. An avid rider, he felt that no hour of life is wasted that is spent in the saddle. Although he rarely went to sleep before the early-morning hours, he insisted that he never be awakened for news except in the most dire circumstances, even though his country was at war, its very existence in peril.

Sounds like a slacker. But this man was not a slacker by any means. This man was Winston Churchill, and he was seventy years old when he led his country to victory against overwhelming odds. His easy mix of leisure and work, as well as his comfortable if idiosyncratic custom of greeting dignitaries in his bathrobe, helped him not only be an effective prime minister of England but also to win the Nobel Prize for numerous literary works, including his definitive six-volume history of World War II. The British war slogan "Keep Calm and Carry On" seemed to mirror Churchill's philosophy.

Churchill was a man of enormous achievement, not the least of which was his skill at catering to his vagal tone.

Although he worked into the wee hours, he was able to harness this productivity to suit his own biorhythms, firmly weaving relaxation into his busy schedule. What Churchill so powerfully conveyed was that sleep-deprived work habits do not demonstrate patriotism or political success, but instead can sabotage the complex decision making at the heart of superior leadership.

Few of us would choose to meet with heads of corporations, much less heads of state, garbed in our pajamas, but we could all stand to take a page from Churchill's workbook. Indeed, a robust body of research suggests that setting aside time for adequate sleep and specific periods of relaxation can actually enhance productivity. In addition, I know of no research that finds that our culture's near fanatical separation of work and family—and our total devotion to a multitasking lifestyle—are anything but detrimental to our productivity and certainly to our sense of calm. The irrational categorizations perpetrated by our rational frontal lobes causes consternation in our core brain. Just being aware of this type of artificial distinction and its effects may help us alter our stress response.

Sadly, modern life vilifies relaxation as much as it promotes rigid work-play divisions. If we were to read that President Obama or any other world leader had issued a directive not to be awakened with important news, we would be horrified. In fact, when Obama and his family vacationed in Hawaii during the BP oil spill, or George Bush went off to Kennebunkport during the Iraq war, they were each pilloried by opponents and even supporters. Yet we know that these healthy retreats from high-octane pressures elevate vagal tone and provide the opportunity for both sociability and inner reflection, which heighten the senses and boost cognitive skills.

I know of a man who stopped having any semblance of a normal life by the time he was two. Now without home or country, he has lived a tumultuous life, surrounded by political intrigue, violence, and unrest. Yet he has continued to maintain an impressive amount of calm. Not only that, he has been able to communicate this feeling to those around him, spreading a contagion of calm. How was the man able to do this? How, when in all likelihood, he experienced exceptionally high levels of stress beginning very early in childhood, did he develop the right balance of his stress response? How does the fourteenth Dalai Lama do all that he does? The answer, in my opinion, lies within his core brain and his vagus, and also within his exceptionally well trained frontal lobes and their capacity for compassion, essential ingredients for brewing calm.

CHAPTER 5

Top-Down Judgment

One year on vacation, I spent several days visiting Will and Sara on their ranch in New Mexico, surrounded by breathtaking views, Angus cattle, horses, dogs, and endless sunshine. Sara told me they both chose to settle here for the same reason, "the absolute peace of this place." But their similarity ended there. Will, a third-generation rancher, had an easygoing laconic way about him, while Sara, an East Coast transplant, always seemed tense, like a coiled spring.

Their distinct temperaments became obvious as I watched how they dealt with a task or a challenge. On my second day at the ranch, I was helping Will build a wire fence strung between posts set a mile apart, amidst vibrant patches of adobe, cactus, and mesquite. We were almost near the other end of the fence when we realized the wire had snapped. I was upset, but Will was unperturbed.

"Well," he said, "reckon the wire's snapped. We're going to have to go find out where it gave."

Under the blazing sun, we retraced our steps to where the wire had broken. Will dug out his pliers and, in his characteristic methodical fashion, expertly linked the edges of the wire together. Then we climbed back into his pickup truck and returned slowly over the bumpy adobe landscape. Suddenly, we heard a snap again.

Annoyed, I said, "Not again! Did we do something wrong?"

"Nope," Will said confidently, "we did nothing wrong. Sometimes with a snap, we get kinks in the wire, and maybe there's a kink in this wire somewhere."

He was right, of course; we found the kink and eventually restrung the wire. It took us more than three hours to finish the chore. Even so, Will remained calm, never becoming angry or irritable. Sara would not have been able to tolerate the fence situation. She would have likely blamed herself for the snafus and for the extra time the job took. While she had "always wanted to be a cowgirl," Sara still had trouble embracing the rhythms of the ranch, constantly racing around and forever trying to get the cattle to conform to her formidable will. As her husband observed one night over dinner, "Sara is taxing, on herself and on others."

Sara said, "I don't know how Will does it! I call him Cool Will, because he's so infuriatingly calm. I like to be prepared, but he says, 'don't fix it if it isn't broken.' Whoa! I think what if it's getting close to broke? Will says I'm a worry wart."

"Sara, no wonder you and Will make such a good couple," I exclaimed. "If he were like you, you would drive each other crazy."

As a neurologist, I found myself observing my hosts through a particular lens. I realized that Will and Sara embody the two ends of the spectrum of frontal-lobe function. The frontal lobes are the seat of control, of choice, of worry, of compassion, of multitasking, and of efficiency. Sara and Will employ each of these qualities in different proportions, making for two distinct personalities, one inherently prone to anxiety, the other disposed to calm.

Will was like a quarter horse, dependable and predictable. He controlled what he could control, didn't worry about things he couldn't do anything about, and put me immediately at ease. Sara was more like a thoroughbred, high-strung and edgy. She made me a little bit anxious. I remember struggling with a barbed-wire gate on the ranch, taking forever to open it. Will watched patiently, not saying a word even as I fumbled with this seemingly simple task. Sara, on the other hand, would give me precise instructions for every chore I undertook, transmitting both her impatience and her anxiety to me. "Do it this way, not that" or "That's too close!" With Sara, I got the job done well and quickly. But Will taught me to reason it out myself as I muddled through under his benevolent gaze.

Sara's ways, while unquestionably more efficient, were harder on her and unhealthy. One morning over coffee, she told me bluntly that she lay awake most nights, worrying about the chores that needed doing, the bills that had to be paid. She ate too much and had put on weight, for which she constantly berated herself. Will, on the other hand, "eats what he wants and never gets fat, and he sleeps like a log," Sara said. "Nothing ever bothers him."

Maybe you want to be more like Sara, but I doubt it. Will's frontal lobes were certainly working in his favor. The question, then, is: How do we become more like Will and less like Sara? The answer lies in better frontal-lobe management. One of the key ways the frontal lobes create a sense of calm is by accurately assessing our ability to control not just our environment but also the people around us. Will does not worry about the calves left behind by the cows, for example, confident in the mothers' abilities to locate their young, and chalking up the one or two that the coyotes get as an unavoidable loss. Sara worries constantly about the coyotes, even though nothing can be done about them.

Creating Order from Chaos

The frontal lobes strive to create order from chaos, to make sense of randomness. It is because of this part of the brain that superstitions persist, that psychics, weather forecasters, and other prognosticators are here to stay. Superstitions impose the illusion of control on events of chance. "My parents were killed by a drunk driver because a black cat crossed my path yesterday." Suddenly, an arbitrary tragedy becomes something under one's control—"If only I could have stopped that cat, my parents would be alive today."

There are more superstitions about avoiding calamities than there are about bringing good fortune, and both economists and fortune-tellers know this. Ever get those chain e-mails that ask you to forward the message on to ten people so as to avoid bad luck or gain good fortune? Which one are you more likely to forward on? If you answered the "good luck" message, you would be among the minority. More people tend to forward chain e-mails that promise to avert disaster, attesting to the risk-aversion role of the frontal lobes.

The psychologist Daniel Kahneman, who won the Nobel Prize for his work in behavioral economics, found that most people avoid risk when they are considering possible gains, but choose risk when they're faced with possible losses. The brain, and the frontal lobes in particular, avoid losses, because loss of status and power are difficult both from the evolutionary and modern-day perspective. Speaking to this "loss of face" is the fact that people will often reject a beneficial offer if they perceive the offer as unfair. Sara is both risk aversive and tries to control the future, while Will, who is also risk aversive, does not try to control that which he is unable to control, allowing him to maintain internal calm. In this function, the frontal lobes mediate between core-brain needs and fears and frontal-lobe rationality and logic. A disconnect between the two makes for unease and anxiety.

The frontal lobes take in current data, add generous dollops of past experience, and cups of anxiety or calm, as the case may be, and spit out forecasts. But despite their attempts at rationally predicting the future, the frontal lobes are poor at doing so; they tend to exaggerate outcomes, both good and bad. This tendency to exaggerate possible consequences may help with the risk aversion that we aim for. If you overstate the negative consequences of a risk, then you're even less likely to take that risk.

If I were to ask myself how happy I'd become if I won the lottery or how sad I'd become if I permanently lost the use of my legs, I'd answer "deliriously happy" to the first question and "devastated" to the second. In fact, the reality is startlingly different. A study to evaluate just this question found that, one year after becoming paraplegic or winning the lottery, both sets of people had returned to their *prior* level of happiness. This seems improbable, even impossible. But in fact it's been found that people generally revert to their baseline level of happiness and calm, regardless of the calamity or the fortune that befalls them. Time evens out the highs and the lows, and regardless of our circumstances, our brains return us to our calm set points. It seems, then, that, as taught in many meditation classes, we synthesize calm regardless of the stresses of our lives.

Our frontal lobes' prediction of the future is partially based on our internalized values. An interesting study a few years ago found that Freud was right all along when he said that during dreaming our critical judgment, or "superego," is suspended. As we dream, the part of the frontal lobes associated with judgment and critical thinking, the dorsolateral prefrontal cortex, is turned down. This frees us, allowing us to fly and to walk on water and to frolic with unicorns. Sleep is essential for calm, and one way it achieves that state is by giving our judging faculties some respite at night. Not judging allows compassion and calm to reign.

Not judging also helps us tamp down our need to control the world around us. "I am a little bit of a control freak," Danny, my architect, told me. "I like everything just so. If things aren't that way I get frantic. It's a good thing I never had kids. I can't imagine their grubby little hands on my art deco table. Maybe it's my army childhood. If dinner isn't served at six o'clock, exactly, I start to freak out.

"Six months ago, I was getting intolerable to be around. My partner and my friends didn't want to be near me. Even my dog would cringe when I came home. What bothered me the most was world events. I would read the papers, watch television, and the more chaos there was out there, the more I would try to control my own personal world. I realized that the worry and fear were making me irrational. So I decided to do something about it."

What Danny did was tune media out of his life. And with this simple act, he went a long way toward freeing himself from anxiety. "I no longer subject myself to these images of suffering playing in my head and messing with my mind. For me, ignorance has been bliss."

Danny's frontal lobes had become frenzied when the

larger world around him spun out of control. Tsunamis, terrorist attacks, brutal crimes "played in his head," tormenting him. He responded by exerting tighter and tighter control over his immediate world, until he became so rigid and demanding that no one wanted to be around him. For Danny, getting off the media treadmill was a viable option for finding calm.

Without our frontal lobes, we would not be able to make the choice Danny made. It took two million years to transform from our ancestral Homo habilis to our Homo sapiens brains, and the brain tripled in size in that time, primarily by the growth of the frontal lobes. In our Homo sapiens brains, the frontal lobes, located on either side and just under the temples, are the pillars of judgment, the arbiters of choice. They give each of us our own individual personality. They are what makes people like Sara persnickety, people like Will mellow, and the Dannys of the world crippled by their obsessions. By mediating between the basic and instinctual needs of the core brain and the needs of society, the frontal lobes help us achieve calm.

The Anxiety Machine

I have used the term "frontal lobes" without focusing on their attendant parts. The frontal lobes responsible for voluntary movement are closer to your vertex (the motor cortex), and all of the frontal lobe in "front" of this narrow strip is called the prefrontal cortex. The parts of the prefrontal cortex closer to your midline and behind your nose are called the orbitofrontal cortex and the medial frontal cortex, and they are older than the part of your frontal lobe behind your temples, which is called the dorsolateral prefrontal cortex. The parts of the frontal lobes closer to the midline are more closely connected to the core brain than the parts that are more to the side. Damage to different parts of the frontal lobes causes different symptoms, ranging from complete apathy to effusive ebullience, attesting to the vital role of the frontal lobes in keeping us socially viable.

The frontal lobes—and particularly the dorsolateral prefrontal cortex—flowered through evolution, growing bigger and bigger from rat to man, to become the largest part of the brain. But the core brain remained essentially unchanged. The frontal lobes control all the lower, evolutionarily older structures, exerting order over our instincts and feelings by imposing rationality and logic. The core brain is a great wild horse. Evolution handed the fragile reins of this tempestuous steed to the frontal lobe, but evolution is an ongoing process, and it is likely that over the next millennia, these reins will become stronger and stronger, making us less "animal" and more "human."

Before evolution gets us there, however, there is a delicate seesaw of negotiations, as the frontal lobes use rational thought to reconcile ancient core-brain feelings. In our ancestors' times, fear, anxiety, and instinct were more crucial for our survival than logic and abstract thinking, so the connections from our core brain up to our frontal lobes are far thicker and stronger than the reverse top-down connections between frontal lobes and core brain. This is why the immediate enjoyment of a hot-fudge sundae overrides the far-off pleasure of looking good in your bikini. Decision-making always involves juggling impulses, generated by the core brain, and thoughts, generated by the frontal lobes. Any asynchrony in this process may contribute to a lack of calm, as seen in the story of Pavlov's dogs.

When the pioneering physiologist Ivan Pavlov paired a painful stimulus with a bell, his dogs became anxious at the sound of the bell even after the pain was removed. And by subsequently pairing a food reward with the bell, he was able to extinguish their learned anxiety. The dogs suppressed their association of the bell with pain and instead associated it with food. However, when water flooded the kennels, they lost the ability to suppress their fear and became frightened once more at the sound of the bell. A stressor, in this case the flood, made the dogs revert to the behavior they had initially learned.

This learning process explains why it is so hard to quit smoking or begin a diet when you are especially stressed. And it is why, in times of extreme stress, we revert to self-soothing skills we learned as children, for those behaviors are associated with a greater level of calm. Although we are no longer likely to suck our thumbs to calm ourselves, we can still choose to bite our nails to the quick. Or, in extreme cases, curl up in a corner and rock ourselves to calm.

As Randolph Nesse explains it, "When you extinguish an anxiety response, you don't undo the learning, you suppress the learning with additional frontal lobe inhibition." So, when we are faced with anxiety, old harmful habits that were overridden by subsequent learning may resurface. This is why, self-destructive as it is, it is not uncommon for heavy smokers dying from lung disease to continue smoking, or even take it up again under the stress of their illness.

This kind of inexplicable behavior speaks to the attempts by the frontal lobe to control us. It makes me think of my friend John, who always joked that he was "hooked" on the news. His fixation reached its peak in 2000, during the Florida ballot debacle that put George W. Bush into the White House. John watched the armored car bearing the ballots slowly wend its way down the highway. During the tense days that followed, he was glued to the TV and totally unavailable to his family. John was so wrapped up in the bizarre turn the election had taken that he couldn't tear himself away from the unfolding drama.

"I feel like something could happen while I'm not watching," John said to his wife, Grace, when she complained about his behavior. "As soon as this is over, I'll do anything you want."

"It doesn't matter," Grace retorted. "Your watching is not

going to change the outcome. You know that! You know you have no control over it."

"I know," John said, "but what if something happened?"

The truth is that we are powerless to influence many of the outcomes in our lives, and the more we try to control events and circumstances, the less calm we inevitably become. It is the frontal lobes that are responsible for this struggle. As our risk-averse frontal lobes, which exaggerate the harm of potential calamities, try to control all that is controllable, we trigger our fear and vigilance systems, creating a body state of constant alert. Unless we renounce some frontal-lobe control, it is impossible to achieve any level of calm. This kind of detachment is what yogis and meditators aim for, as they surrender control. Experienced mediators are able to ward off their internal, quasi-rational frontal-lobe demons without changing anything outside themselves. This ability leads to an inner sense of stillness.

Cultivating Compassion

The skillful management of the frontal lobes brings to mind a man I know who, like Will, projects calm and confidence to the people around him. Moses is a beloved doorman of a large New York City apartment building—he calls himself the "mayor of the block." In the thirty-five years that he has manned the door, he has always greeted everyone, friend and stranger alike, with a ready, happy smile. He takes a real interest in their lives. And that interest and the empathy that underlies it have inspired residents to confide in Moses, a man who did not go past grade school. Lawyers and doctors and corporate executives seek Moses's counsel. Teenage girls cry on his shoulder, complaining about their overbearing parents, and the said parents vent their frustrations about their headstrong daughters. Moses always has the same response, which he delivers with calm conviction. He says, "Now, don't you worry. The good Lord will take care of it." Even though he is deeply religious and many of the residents are not, he communicates his reassurance and helps soothe their worries.

What is it about Moses's religion that imbues him with calm? I believe that he and others of deep faith are able to deactivate the judgmental, questioning parts of their frontal lobes. Compassion, so evident in Moses, is the ability to stop questioning and judging and to become empathic.

In recent years, researchers have attempted to evaluate the effects on the brain of compassion and faith. In one study, it was demonstrated that Christian subjects deactivated the medial and dorsolateral prefrontal cortices, the parts of the frontal lobes involved in critical thinking, when they listened to speakers who they believed had healing powers. The level of deactivation depended on the strength of the participants' religious beliefs and how charismatic the speakers were, as well as on how deeply the participants felt God's loving presence during the session. The term "blind faith" may not be far off the mark. People who are able to immerse themselves profoundly in a religious experience often suspend disbelief and rational thinking, which leads to a feeling of calm.

Great leaders are able to create the same calming effect. Franklin Delano Roosevelt famously soothed a distraught nation by resoundingly declaring, "We have nothing to fear but fear itself." The virgin Queen Elizabeth mobilized her disheartened troops by riding out in armor as their teenage commander-in-chief and spurring them on to unprecedented victory.

Judging yourself and the people around you is part of the frontal lobes' normal function. But being overly judgmental, setting standards that are difficult to attain, does not lead to compassion or calm. Compassion, not just for others but also for oneself, is intrinsic to a sense of calm. Will didn't blame himself when the wire broke twice, and he was tolerant and reassuring when I struggled with the simple task of closing the pasture gate. But Sara seemed incapable of forgiving herself or anyone else for mistakes or shortcomings, and this lack of compassion added to her stress.

Several studies have demonstrated that self-judgment and compassion reside within the frontal lobes, producing opposite effects. Self-critical thinking is strongly correlated with a variety of anxiety and mood disorders, while compassion is associated with health and calm. In one experiment, seventeen young adult women volunteers were presented with two scenarios while undergoing imaging to measure brain activation. An event representing a personal setback or mistake, such as "A third job-rejection letter in a row arrives in the mail," was contrasted with neutral occurrences, such as "The second free local newspaper in a row arrives in your mailbox." The young women were then asked to respond to each negative scenario with either self-reassuring or self-critical thoughts. Self-criticism was associated with activation of the dorsolateral prefrontal cortex, the judgmental and critical area of the frontal lobes. In addition, those women who rated themselves as highly self-critical were more likely to have higher activity in this part of the frontal lobes to begin with. This area appears to be associated with a preoccupation with mistakes and failures.

Despite this apparent hardwiring, there is some evidence that compassion and self-reassurance can be learned. An imaging study that examined both trained monks and volunteers who had only brief experience with meditation found that activity in the parts of the frontal lobes dealing with self-judgment was reduced in both groups. The novice meditators had apparently achieved enough facility in the few weeks they were trained that the results were apparent on a brain scan. In another study, in which volunteers took a meditation course, learning to focus on the present moment, the parts of the frontal lobe that deal with compassion were more activated than in a control group that did not take the course.

It is striking that Buddha reportedly identified both

inner calm, *samada* in Sanskrit, and insight as the ultimate results of a productive meditation. But the meditative traditions are not the only way to boost one's compassion and reduce self-judgments. In fact, popular therapeutic treatments like cognitive behavioral therapy have been shown to be effective primarily because they methodically teach clients to reject self-critical thinking and to embrace their positive qualities.

Another function of the frontal lobes is to interpret the experience of pain. Interestingly, though, the frontal lobes perceive disconnection from others—social pain—as equivalent to physical pain. In an imaging study of college students, their frontal lobes lit up both when painful pressure was applied to the students' nail beds and when they were presented with a scenario of social rejection.

Both bodily and emotional suffering diminish compassion and compromise calm. So does exposure to violence, including watching violent images. This was what happened to Danny before he decided to detach himself from the news media. But bonding with others enhances compassion and helps produce calm. Sometimes it can even create what seem like neurologic miracles. I saw this dramatically demonstrated in my patient Stella, a sixty-two-year-old grandmother who had been married for four decades to Al, the "man of her dreams."

One morning, Stella started to lean over in bed to kiss Al and found she couldn't move at all. The right side of her body was paralyzed. Terrified, she tried to call out to Al but discovered she could only grunt with great effort. Stella managed to shake her husband awake with her left hand, which was still moving.

In the emergency room a few hours later, Stella attempted to follow my directions as I examined her. By this time, it was clear that she had suffered a stroke. Her face was grotesquely twisted to the left; saliva leaked out of the flaccid corner of the right side of her mouth.

"Show me your teeth," I said, as part of my assessment of whether she could understand and follow spoken commands.

Al looked on in shock as the left side of Stella's mouth retracted, revealing an even row of dental implants, but the right side lay unmoving in a strange pout.

"Lift your arms," I said.

Al watched his wife of forty years struggle to follow my instructions. He stared in horror at her right arm lying inert at her side like a dead fish, even as her left arm obediently rose. Tears welled up in Stella's eyes and rolled down her cheeks, soundlessly salting the sheets. The sight of his usually spunky Stella was more than Al could bear.

"You keep that crying up and I'm gonna smack you so hard, you won't be able to move your other side either!" Al said in a mock threat.

At this, Stella started laughing. And Al stared at her in amazement. Because when she laughed, Stella's face was no longer paralyzed; the sides of her mouth moved symmetrically in the smile that he knew and loved.

How did this happen? How could Stella's face be paralyzed one minute and move normally the next? The answer is simple. Her frontal lobes understood Al's mock-threatening joke. They bypassed her voluntary circuits and triggered her core-brain circuit, getting her to bond with Al. The frontal lobes are responsible for our ability to fake emotions, a facility we share with chimpanzees but not with lower animals. Because Stella's stroke destroyed some of her frontal lobe, she was not able to fake a grin on my command. But when Al made a joke, Stella's core brain, with its lower-animal mechanism for displaying emotion, enabled her to smile. True emotion triumphed over facial paralysis.

This type of emotional bonding—the powerful connection between Al and Stella—suggests why the genuine and spontaneous smiles of little children can have a soothing effect on adults. Their expressions of pleasure and joy invite us to bond with them and thus to experience the emotional connections that lead to calm.

Stella exemplifies another important function of the frontal lobes, the ability to relate to other people so as to maintain a sense of community. Compassion, empathy, and judgment of self and others are frontal-lobe roles intended to connect us with each other. People who do not react appropriately to emotional cues—such as those with Asperger's syndrome—cannot fully flourish in society, regardless of talent or intellect. Once again, evolution is at work. Your frontal lobes allow you to laugh heartily at your boss's lame joke, thereby securing that promotion and a better future for your children. They disguise the confusion on your face when you are asked an unfamiliar question during an interview, giving you time to come up with a plausible answer. They mask your disgust with a smile when your toddler presents you with the headless mouse he found on the street.

The frontal lobes thus foster calm by helping with the nuances of social interaction—and deception—so important for living in groups. Poor frontal-lobe function contributes to the distress that people with autism experience when they're confronted with change, or the anxiety those with obsessive-compulsive disorder feel when their routines are interrupted. These reactions also lead to social isolation, exacerbating anxiety, and diminishing calm.

We can think of imitation and the connections it fosters as the opposite of social isolation. Imitation is pervasive and automatic in humans and is thought to be a function of so-called mirror neurons in the brain. There are brain areas devoted solely to the mirroring and imitation of other persons, just as there are brain areas devoted exclusively to monitoring other people's faces and emotions. The extensive representation of these functions in the human brain speak to their importance for our survival. Mirror neuron theorists postulate that if we didn't imitate other people, we would literally have to reinvent the wheel every generation and would not be able to build on past knowledge. There would be no Facebook and no Ferrari. Mirror neurons have been critical to human civilization.

The discovery of mirror neurons in the 1990s added a significant dimension to our understanding of imitation and of the part it plays in social interaction. For example, studies have shown that children with autism—who typically find it near impossible to interact socially—are deficient in mirror neurons. An imaging study that compared imitation and observation of facial expressions in children with autism and in normal children demonstrated not only a deficit in mirror neuron areas in the children with autism but also a correlation between the severity of the disease and activity in these areas: The lower the activity in mirror neuron areas, the more severe the autism.

Mirror neurons would not exist if we were meant for a solitary life. Neuroscientist Marco Iacoboni, whose group first identified mirror neurons, believes they provide an unreflective "inner imitation" of the facial expressions and actions of other people, and this imitation does not require conscious, deliberate recognition of the expression mimicked. In other words, observing someone's raised hand will trigger a mirroring action within your own brain cells, as if you, too, were raising your hand. Unconsciously, your brain determines that the hand was raised to answer a question in class rather than elevated as protection from falling debris. At the same time, mirror neurons send signals to the core brain, and core-brain triggers then allow us to feel the emotions associated with the observed action. Only *after* we feel them emotionally are we able to explicitly and consciously reconcile the nervousness about the falling debris with the excitement of knowing the right answer in class.

If this is true, there should be a relationship between the tendency to imitate others and the ability to empathize with them, which increases the likelihood that two people will feel affection or affinity for each other. Imitation is, indeed, the best form of flattery. The more you mirror another person physically, the more you begin to care about the feelings of that other person, the more empathy you have. In one study, the more a subject mimicked another person, the higher the subject scored on empathy measures. Through imitation and mimicry, even of simple physical acts like foot shaking or nose rubbing, we become more sensitive to complex emotions. And this allows us to respond compassionately to another person's emotional state.

Our frontal lobes sensitize us to those around us. In a study of ageism, subjects who were exposed to words associated with the elderly—such as *Florida*, *bingo*, and *gray*—walked more slowly and remembered less than subjects who were presented with age-neutral words. Thanks to our mirror neurons, when we are around older people, we are able to empathize with their way of being. When we are around calm people, we become calmer. Interestingly, it is well known in disaster management that people seek safety within crowds, even when it is safer to be alone. We feel calmer and less anxious when we are with other human beings because of the empathy our mirror neurons generate.

Empathy is one of doorman Moses's strong suits. He knows instinctively how to establish powerful personal connections. When he chats with a CEO, he folds his arms and keeps a distance and nods his head slightly. When he speaks to a little girl about her lost toy, he crouches next to her and mimics her body language. By mirroring actions, Moses conveys his deep empathy and maintains his calm brain.

Juggling Memories

Memory processing—mentally juggling vast amounts of information—is another major frontal-lobe function. Unfortunately, though, this skill has made the frontal lobe particularly suited to the modern demands of multitasking, which is a major impediment to calm. The frontal lobes are important for their ability to focus and to resist the intrusion of irrelevant or distracting information. When this process is inhibited, a person is unable to maintain focus, moving from one action or object to another. Too much of this mental "flitting about" causes problems like attention deficit disorder. But too much focus makes Jack a very dull boy.

The extreme form of focus and inability to switch tracks

is called perseveration and often occurs in frontal-lobe damage. There is a fine line between perseveration and perseverance. Perseverance is a valued trait without which we'd scarcely be able to achieve goals. But it can take on pathological extremes. In conditions like Alzheimer's disease, for example, perseveration is common. Often when I ask these patients to stick out their tongue, they do just fine. But then when I switch tracks, they get derailed. "Raise your right arm" elicits another obedient tongue display.

Perseveration underlies the kind of ruminative thinking that is at the heart of anxiety. It is the antithesis of calm. Sara lies awake at night, ruminating, stuck in the rut of her worries and deadened to the life-enhancing possibilities of change. "The only difference between a rut and a grave are the dimensions," the Pulitzer Prizewinning novelist Ellen Glasgow famously quipped.

Sara is a multitasker, Will a serial tasker. Serial tasking allows Will to luxuriate in his early mornings on horseback, the demands of the ranch held at bay. This does not make him less efficient, just less frenzied and more calm than Sara, who is mentally miles away from the mesas where she has moved to find peace.

Cultivating serial tasking not only reduces your false sense of urgency and helps you relish the moment; it allows you to listen more sensitively to others. I was able to connect to Will because of his attentive listening, and this was calming for both of us. Experienced meditators are expert serial taskers. They sit still, relinquish control, and focus on just one thing. Their vagal tone is increased as they modulate their breathing, their heart rate is slowed, and they reach a point where their core brain and frontal lobes are as close to being in sync as possible. They become calmer.

Technology exacerbates multitasking problems. Even though staying online with your iPhone makes it possible for you to sneak out of the office to watch your son's soccer game, you are still hooked to your job and not entirely present for your son's triumphs on the field. Every time you decide not to look at your BlackBerry, you're making a choice mediated by your frontal lobes. This creates an undercurrent of vigilance that does not allow you to relax and enjoy the game.

Given their skill at multitasking, it is no wonder that the frontal lobes excel at monitoring and measuring time—and making sure we are aware of it every waking moment. Twitter, smartphones, and other technological advances—with their instant updates and constant feedback—accentuate the sense of the inexorable passage of time. The productive imperative to accomplish everything in as little time as possible contributes to anxious feelings, even as we pat ourselves on the back for our achievements. In Eastern traditions and among other religious devotees, there is a continuity of time, an awareness of time being cyclical rather than linear. In Western culture and in urban environments, there is a sense of time running out. The frontal lobes create urgency when none is needed, alarming us and propelling us into a state of anxiety.

The Trouble with Freedom of Choice

Making choices is also the job of the frontal lobes. The psychologist Barry Schwartz, author of *The Paradox of Choice,* points out that while choice maximizes individual freedom, it also shifts responsibility to the individual. In fatalistic or religious societies, there is less stress even with severe hardships because there is less personal blame, more personal compassion, and less sense of personal choice. In these societies, the frontal lobes are not as troubled by the stress of choice. Schwartz notes that too much choice has two negative effects: It can cause paralysis and it can bring on discomfiture. We are less satisfied with our decisions than if we had fewer choices. Moreover, people tend to suffer greater regret after making errors of commission—as in so-called "buyers' regret"—than after errors of omission. By not choosing at all, you are paradoxically less likely to rue the de facto choice.

My friend June can testify to the "paradox" of choice. She was recently shopping in a department store during a major post-holiday sale. She told me what happened.

"I suddenly had an overwhelming sense of panic. All the clothes I wanted were on sale. The colors I like, the right size. I remember staggering to and from the dressing room with masses of pants and tops and dresses clutched to my chest. I started trying on bunches of clothes, and all of a sudden in the middle of it, I felt overwhelmed and terribly anxious. I've never had that feeling before or since, but, I tell you, it was terrible."

In the face of limitless choices, June had succumbed to anxiety—and wound up buying nothing despite the terrific array of clothing at great prices. The frontal lobes, which are constantly weighing the pros and cons of every bit of information, look at choices rationally, trying to determine whether the price is right. But the right price doesn't necessarily have to be a low price; sometimes a high price is more tempting to the frontal lobes, which aspire to this snob appeal.

The frontal lobes also set up a "what if" expectation, as in "What if my choice is not the right choice? What if the boots I picked are not really the best boots? What if the person I choose to marry is not the right person?" In days gone by, there were far fewer consumer goods from which to choose. Our spouses were picked from a select group of people, perhaps from the same village, and there wasn't much choice involved. So there was little second-guessing.

The frontal lobes are the masters of the second guess. Indeed, for the frontal lobes, choosing means losing other options. You can never, ever, be truly content, truly calm, with a choice, simply because there's always the possibility of a better choice. The core brain operates instinctively. It does not have the wherewithal to second-guess this way. Something feels squishy to your core brain, it's happy with it. Something feels slimy to your core brain, it doesn't like it, even as the frontal lobes advise that the tripe soup in your bowl is really a delicacy. Here again is the discord, the frontal lobes trying to ignore core-brain choices and overriding core-brain needs. Blotting out those needs doesn't result in calm.

The journalist Malcolm Gladwell, in examining consumer choice, writes that polling people in focus groups is not the best way to figure out what they want. He quotes Howard Moskowitz, who introduced us to extra-chunky spaghetti sauce, catapulting a struggling Prego past its rival, Ragu, into marketing history. Moskowitz observed: "The mind knows not what the tongue wants." To put this in neuroscience language, the core brain (tongue) determines taste preferences; the frontal lobes, which flourish cheerfully in focus groups, have little idea of what the tongue is up to.

What if two tastes appeal equally to the core brain? Then, as demonstrated by "elite" Grey Poupon's successful advertising campaign against ordinary mustards like French's, the marketers have to appeal to the resident snobbery of the frontal lobes. One way to make people happy—and thus persuade them to buy your product—is to give their frontal lobes something they aspire to. When the frontal lobes focus on an achievable goal—such as acquiring the mansions, butlers, and limousines associated with Grey Poupon—calmness holds sway.

Finally, lest you assume that the frontal lobes are completely under our control, here's something to think about.

The concept of free will has been debated by philosophers and theologists. Even so, most of us would agree that nothing could be more deliberate and under one's control than one's own movement. Nothing can be more under my control than the willed movement of my hand. And yet . . .

In 1965, German scientists described the extravagantly named Bereitschaftspotential—or readiness potential—a change in brain activity that appeared a whole second (!) *before* a voluntary movement. This may not seem like much time, but understand that most activity in the brain occurs in the order of milliseconds. In a recent experiment at UCLA, twelve patients with epilepsy, who had brain electrodes implanted for seizure control, were asked to watch an analog clock on a laptop computer and to push a button after at least one rotation of the clock's hands whenever they felt the "urge to do so." Each time they did so, they were also asked to estimate where the clock hand had been when they felt the urge. By evaluating the firing of just 512 nerve cells within the frontal lobes, the researchers were able to predict with a 90 percent accuracy when the patients first "felt the urge" nearly 500 milliseconds (or nearly half a second) prior to the patients' actions. In other words, there is no free will, even with the frontal lobes.

So for all their foibles, the frontal lobes are crucial to the dance of a calm brain.

Let us now turn to some of the intermediate players, the structures that connect our frontal lobes and our core brain. These not-so-bit-players are a step down from the rationality of the frontal lobes and its top-echelon judgments, but a step up from the sheer animal instincts of the core brain. Some people consider them a part of the core brain. But I like to think of them as mediators, negotiating at the gates between our senses and our sensibilities.

CHAPTER 6

Gatekeepers

The frontal lobes are the jewels of human evolution, at the root of all that is modern life—nuclear weapons, smoke detectors, supercomputers, and survival in our new world. But the old world of our core brain—which is responsible for falling in love, for saving you from predators, and for keeping your grandparents alive so that you can now read about calm—continues to be more powerful, navigating the challenges of the twenty-first century, using neural tools developed before the Iron Age. In reconciling these two willful structures, one evolving in a world without nuance and the other in a world full of nuances, what I am calling the gatekeepers are as crucial as the frontal lobes.

These key structures—the amygdala, the insula, the subgenual anterior cingulate, and the nucleus accumbens—are clumps of tens of millions of nerve cells organized into fairly distinct functional entities that connect the core brain to the frontal lobes. They are located between the stem of the core brain and the sprouting and spreading of the overlying new brain, including the frontal lobes. These gatekeepers are sentries in this path, letting in some information, blocking other information, and in some instances, deciding to act on their own without ferrying the information further up into our consciousness.

Here's the kind of thing these gatekeepers do: I am hiking Acadia National Park, lost in the wonder of the nature around me. I am about to step over what looks like a twig. My core brain sends up information to one of my bit players, the amygdala, and before I am consciously aware of it, I step away from the twig that looks like a snake. By the time I consciously realize what's occurred, I am milliseconds away from harm—a critical chunk of time for survival.

In Holland, scientists watched in amazement as a physician who had suffered strokes that wiped out all of his vision navigated a corridor strewn with obstacles, including a trash can and a ream of paper, without bumping into anything. Even the most staid scientists burst into applause at the end of this seemingly miraculous feat. How did he do it? Although his new brain couldn't consciously see and register the objects in his path, his core brain and his gatekeepers processed visual information from his retina appropriately, keeping him from injuring himself.

Optimal interaction among these structures leads to the

chorus of calm, beginning in our bodies, channeled up by the vagus, the parasympathetic and sympathetic systems, modified at the core-brain level, and sent further up into our awareness. As we've seen, the connections from the top down are more tenuous and tendril-like than the bottom-up cable-width links. With this lopsided wiring, the bottom-up well-spring of the body's needs trumps the top-down trickle of rational and conscious desires. This dialogue flow contrasts with the modern top-heavy approach to calm. The gate-keepers are the way stations along this road, processing much of the information locally and letting what they deem relevant go up to newer brain areas.

So, if I am deep in conversation with my hiking partner, I may trudge along without even consciously registering that I narrowly escaped a copperhead snake bite. My core brain and the bit players saved my life without my having to think about it, but they did send on some of the informa-tion after the fact to the frontal lobes, which chose to ignore it, absorbed instead in the discussion of tent suppliers.

Even if he were being attentive, the blind physician still could not see, for he had no newer brain area to handle the information that his core brain and gatekeepers were send-ing up. Even so, the gatekeepers sent down enough infor-mation to his body to steer him clear of obstacles. This same man could recognize faces as angry or sad or happy when flashed before him on a computer monitor, but he could not distinguish a circle from a square. This speaks to the emphasis of the core brain and our gatekeepers on people and community, and why understanding emotions is more important for survival than "seeing."

Neuroscientist Antonio Damasio notes the significance of some of these gatekeepers and their influence on how we make decisions; he points out that they improve decision making precisely *because* they temper logic with emotion. In one of his experiments, he asked patients to choose between four decks of cards: two of the decks yielded high immediate profit but larger long-term loss, and two yielded lower immediate gain but greater long-term gain. After some play, normal subjects avoided the bad decks, with the larger long-term loss, and changes were generated in their skin-conductance response—a measure of the body's awareness of a problem—*before* they became conscious of their decision. Patients with damage to the gatekeeper amygdala, or with the closely allied more midline frontal lobes, showed no such fear of the bad decks. These patients did not learn to choose the more advantageous decks. They also showed no changes in body state signifying anxiety when choosing the bad decks, as did patients with functioning gatekeepers.

It isn't just about snakes. Emotion, mediated by the bit players and the body, helps our frontal lobes make all kinds of better decisions. The "thought" of getting fired, for example, generates a somatic response, which forces you to roll out of bed and get to work on time. This body-based emotion points you in the right direction so that you make the right decision. On the other hand, suppose you have an argument with your spouse in the morning before you get up. Your visceral emotion does not necessarily bias you toward the "right" or most advantageous decision. It gets complicated. You are late for work and may lose your job. The gatekeepers regulate the emotional inputs that help us with decision making, integrating reasoned thought with unconscious impulses from the body.

There are several important gatekeepers, some of which involve the more midline frontal lobe structures. Let's focus on the four I mentioned above that are particularly important to the neurobiology of calm. These are the *amygdala* (fear center), the *insula* (gut-feeling center), the *subgenual anterior cingulate cortex* (sadness center), and the *nucleus accumbens* (happiness center). There are two of each of these structures in our brains, just as we have two arms and two legs. Nature seems committed to symmetry. Taken together, these structures, and several other brain regions that deal with emotion and memory, are referred to as the *limbic system*. A monkey high on a psychedelic drug opened a window into the workings of the fear center.

The Monkey and the Fear Center

The year? 1936. The name of the rhesus monkey? Aurora. Scientists Heinrich Klüver and Paul Bucy, members of the Chicago Neurology Club, were studying the seizures caused by the cactus-derived psychedelic drug mescaline, or peyote, on the brain of rhesus monkeys. Young Aurora, unluckily for her, was chosen to participate in the study and developed seizures after being given mescaline. Klüver and Bucy decided to lop off Aurora's temporal lobes in an attempt to contain her seizures. This part of the brain, which rides shotgun, high and behind the nose on either side, has hidden deep within it not only the amygdala, but also the center for memory.

It turned out that removing Aurora's amygdala failed to eliminate the mescaline-induced seizures, but the operation did change her behavior dramatically. Poor Aurora wandered through the lab for months after losing her amygdala to advance science. She and the eighteen other monkeys that joined her in this temporal-lobeless state were wild and aggressive before surgery; afterward, they lost all fear, approaching a large bull snake or a strange person without hesitation. In dramatic contrast to their presurgical behavior, they became hypersexual—masturbating excessively, copulating continuously for up to a half hour, lifting another animal by its penis while the other animal did nothing more than grunt. Because removal of her temporal lobes also obliterated her memory, Aurora explored with her mouth every object she encountered, as if she'd never seen any of them before. She discarded all inedible objects, such as a nail, a glass, a live mouse, feces, or a piece of sealing wax, and consumed all the edible ones immediately.

Klüver and Bucy succeeded in establishing the location of both the fear and the memory center in the brain, at least in rhesus monkeys. Other researchers found that stimulating the amygdala in a cat produced paroxysmal rage, with automatic snarling, arching of its back, and its fur standing on end. From this and other research, it became clear that the amygdala is involved in both extreme aggression and fear.

Nearly twenty years later, in Padua, Italy, neurosurgeons Hrayr Terzian and Giuseppe Ore described what happens when the temporal lobes are removed from a human being. The patient was a nineteen-year-old man of normal intelligence—let's call him Leonardo—who had epilepsy. In addition to his epilepsy, Leonardo had become progressively more aggressive, trying to "strangle his mother or crush his younger brother under his feet." Aggressive behavior is sometimes seen in patients with seizures arising in the temporal lobes. At the hospital, Leonardo attacked the nurses and tried to commit suicide, although in between these spells, he behaved appropriately.

Terzian and Ore knew of Klüver and Bucy's treating mescaline-induced seizures by removing the temporal lobe, which produced calm in Aurora and her fellow monkeys. They knew of the cat experiments in which stimulation of the amygdala caused rage reactions. They decided that removing Leonardo's temporal lobes might well help him not only with his seizures but also with his aggression. So, on a fine October day in 1952, they excised Leonardo's left temporal lobe. When this didn't have the desired effect, they went on to remove his right temporal lobe a month later. Now would Leonardo behave?

Behave, Leonardo did, although more like Aurora than the mild-tempered young man his mother apparently yearned for. Before the operations, Leonardo displayed an affection for his mother that she "herself considered exaggerated." Now Leonardo called her madam and didn't appear to recognize her at all. A few days after the second operation, his appetite increased so much that he ate "at least as much as four normal persons." And, not unpredictably, Leonardo, like Aurora, lost all his sense of anger. He became meek and childish, and no amount of provocation could elicit aggressiveness or fear. He lost all ability to display emotion and only smiled upon seeing himself in the mirror. The researchers tell us, "He liked to look at himself for many hours in the mirror." Leonardo lost nearly all his memory. Two years later, he was placed in a mental hospital, where he practiced "self-abuse"—mid-century speak for masturbation—several times a day but showed no sexual interest "either toward the male or female sex." Terzian and Ore named the eponymous syndrome when they published Leonardo's story in *Neurology* under the title "The Syndrome of Klüver and Bucy—Reproduced in Man by the Bilateral Removal of the Temporal Lobes."

The sad stories of Aurora and Leonardo demonstrate the importance of the amygdala in managing not only fear and aggression but also social interactions. The amygdala is fear central and our brain's sensitive smoke detector, sending out an all-system alert at any hint of danger. A highly sensitive detector was a crucial part of our ancestors' survival kit. For them, it was critical to move quickly from sensing smoke to thinking fire, as they spotted and then fled from predators. It is no surprise that they sometimes made mistakes—and the predator turned out to be nothing more than the wind in the woods. But, as Dr. Randolph Nesse observes, "When the cost of expressing an all-or-none defense is low compared to the potential harm it protects against, the optimal system will express many false alarms." In other words, the cost of many false alarms shrank in comparison to the expense of making that one mistake that could have cost our forebearers their lives.

For some people—say, executives scrambling up the corporate ladder, young people applying to college, actors auditioning for parts—the amygdala's smoke alarm never shuts off; it senses danger at every corner. Of course, what's at stake is loss of pride or promotion, not loss of life, and in overreacting, the modern amygdala can contribute to omnipresent anxiety and lack of calm. This hypervigilance causes the fear circuits to become overactive and underactivates circuits in the frontal lobe that might have tamped down the runaway anxiety.

An overactive amygdala interferes with laying down memories and retrieving them, and it also sears emotionally laden memories into the structure of the brain. So, for example, if I am overly anxious in an interview, I may have trouble remembering the interview later or I may play back embarrassments within it over and over again. The stress and anxiety, plus the resulting rise in cortisol level, make my smoke alarm go rogue. After some time, if I choose to continue to dwell on the interview's humiliations, irrelevant details get smeared with the same tar brush of fear. So, not just the "smoke"—the memory of the interview itself—sets off the alarm, but perhaps even the color blue (which was the color of the interviewer's dress) or the sound of Beethoven's "Für Elise" (which played faintly in the background). A feature of chronic anxiety is that the amygdala casts a net way beyond the original stimuli, triggering the alarm even when the connection is not apparent.

Let's say a woman is assaulted by a man in a green shirt in the back of a bakery, and in turn the memory of this trauma is branded into her brain by the amygdala. In addition, the amygdala may strengthen the memory center's net for the memory, flinging it over more areas of the brain, using the strength of the emotion to imprint the memory into parts of the brain that deal with vision and smell. Soon a glimpse of a green lawn or the smell of baking bread makes her anxious, although she may not understand why.

Smell is the most primitive and evolutionarily oldest of our senses and is tightly woven into the fabric of the amygdala. Taste is also a primitive sense and linked closely to smell. When patients of mine who have suffered a head injury complain they can't taste anything, they really are complaining of a lack of smell. So when a bite of a madeleine evoked Proust's remembrance of things past, his amygdala is what high school English students have to blame.

When we are exposed to daily anxiety-producing situations, structural changes in the amygdala keep us in a state of fear and foreboding. Fortunately, the amygdala receives input from the vagus nerve to help it gauge threat levels. A high vagal tone reduces the amygdala's sense of danger, downgrading it from high alert and increasing one's sense of calm.

The Monk's Gut Feelings

Gatekeeper number 2, the insula, answers the question "How do you feel?" according to Bud Craig, a distinguished neuroscientist. Have you ever felt "sick to your stomach" because you are anxious? Or anxious and ill because you have the stomach flu? This is because your vagus ferries information up from your body into your insula, and this gut-feeling center processes the information and gives you the queasiness and bellyache that signal anxiety. In contrast, being calm makes you feel "good to the core," imbued with a bodily sense of well-being.

When we don't feel well, regardless of whether the cause is psychological or physical, the insula is involved. In an interesting study addressing this point, researchers measured brain activity after injecting sixteen healthy male volunteers with typhoid vaccine, which makes you feel poorly, or a saline injection which doesn't have any ill effect. After the typhoid injection, volunteers felt "ill," with more fatigue and concomitant activation of the insula. All things vagus end in the insula and let us know about the "State of the Body." Whenever we feel bodily ill, whether from a typhoid shot or because of nervousness, our insula becomes active.

I'm often struck by how many idioms are grounded in truth. And this is particularly true of the insula. The insula lights up when gauging trustworthiness of faces, for example. "I can't explain it. It just doesn't *feel* right!" speaks to the dissonance between what we're feeling at the insular level, our gut feeling, and what we are consciously processing at the frontal-lobe level. Antonio Damasio speaks of the state of interoception, or the knowledge of and perception of the internal body. He observes that the insula is the place where we integrate how we are feeling internally and give those feelings voice. The insula is key not just in integrating our feelings about our own selves but also in integrating our feelings for others, a crucial core-brain function.

The psychologist Richard Davidson has investigated this feeling for others by studying the effects of meditation on the brain. He evaluated the effects of compassion meditation in sixteen Tibetan monks who had trained for years and sixteen volunteers who had had some preliminary training in meditation. He found that when these individuals were asked to focus on feelings of compassion for self and others while listening to various sounds of human emotion, the trained monks had higher activity in the insula, and those who reported more compassion had more activity in the insula.

You don't have to be a Tibetan monk to increase insular activation and a sense of well-being. In another study, half the volunteers took part in a meditation course, learning to focus on the present moment, while the other half did not. Here as well, researchers found that the insula was more activated in the group that had had meditation training. Expert meditators tend to engage consciously in compassion for others and for themselves, and the insular activation reflects this. Complex positive emotions like pride and joy also activate the insula.

This representation of the body's state within the brain sometimes makes us behave in ways that don't always make sense to us. Your friend is late for lunch—again. She has a valid excuse—again. Rationally, her story sounds plausible and you want to be understanding. Again. And yet you say to yourself, "I don't feel good about this." You can't quite enjoy lunch. That tight knot in your stomach is your body's rendition of your angry brain.

The insula informs us in a preconscious way about how we feel. Imaging studies have shown anger to light up the insula, even as the person experiencing this emotion has no conscious clue of it. In other words, sensations in the insula precede any realization and identification of emotion. For calm to reign within your brain, there has to be bottom-up calm from your body, which is carried out by the vagus. The vagus not only carries information up to the fear center and the gut-feeling center, it also carries it up to the middle parts of the frontal lobe, where the brain's sadness center, the subgenual anterior cingulate cortex (also known as Brodmann's Area 25), resides. It is our Gatekeeper number 3. Sadness and depression often seem just around the corner as our anxiety level rises and all sense of calm dissolves.

Checking into the Sadness Center

Helen Mayberg is a neuroscientist who has pioneered in treating depression, using deep-brain stimulation by means of electrodes implanted in the brain. Her group describes what happened to a thirty-six-year-old woman whose sadness center was stimulated during surgery. Since age twelve, this woman had a history of depression that had not responded to medication. When doctors used deep-brain stimulation to activate her sadness center, she responded powerfully within thirty seconds. She became panicky, tearful, and very uncomfortable and was able to tolerate the experience for only minutes at a time. She reported that, during the stimulation, she "did not care about myself or anything. . . . Someone could come in to shoot me and I could have cared less. . . . Deep down bad feeling in the pit of my stomach. . . . Similar in some respects to my depression but a thousand times worse. . . . I wanted to cry but couldn't." As soon as the stimulation was stopped, her mood returned to baseline, but it could be reproduced whenever the stimulation was repeated over the following week.

Depending on the rate of stimulation, one can either increase activity in the area where the electrode terminates in the brain or reduce activity in that area. Thus, one can either cause symptoms of depression or cure it. Mayberg and her colleagues have used the knowledge gained from this kind of brain stimulation to treat severely depressed patients. In four out of six patients who had severe major depression that had not responded to four different antidepressant treatments, including electroconvulsive therapy, there was dramatic improvement that persisted six months after surgery, with an effective cure in three of the patients.

Brain imaging studies show reduced activity in the sadness center with successful treatment not just with antidepressants but also with placebos or chemically inert "sugar pills." In other words, when we *feel* happier, our sadness center is turned down, even if the cause is a sugar pill. The reverse is just as true. Even when someone is very sad but not clinically depressed, there is increased activity in the sadness center. Finally, patients who do not respond to antidepressant treatment show no reduction in activity of the sadness center.

The sadness center is important in regulating negative mood in both depressed patients and normal persons, with both acute and chronic stress. Knowing that the sadness center is malleable in these ways can help us turn down the activity there. Calmness reduces activity of this center.

And, would you believe, so does acting happy regardless of what you are actually feeling. Research has shown the power of "putting on a happy face" by demonstrating that even a forced smile will lower the activity of the sadness center. Because we know that having sad thoughts, dwelling on unhappiness or misfortune, real or imagined, makes the sadness center more active, thinking happy thoughts will help turn it down. The science of positive psychology abounds with clinical accounts of this. To the point of this book, acting calm is helpful in getting to calm.

The sadness center is not the only area involved in producing depression, just as fear is not mediated in the amygdala alone. The gatekeepers are not islands; they work with other brain regions to effect changes in affect. The amygdala, for example, interacts with several parts of the frontal lobes when dealing with emotion. However, for the purposes of simplicity and to prevent everyone from nodding off, I have kept things simple. One neuroscientific fact that is worth mentioning, though, is that overactivity of the right side of the brain causes depression and anxiety, while overactivity of the left side of the brain causes mania. I teach medical students to remember this by saying that *right*eous people are often anxious and depressed, while the *left*-leaning free spirits among us are carefree and happy.

Again, for simplicity's sake, let us set aside distinctions between the right and left brain hemispheres with regard to the gatekeepers and not parse the frontal lobe into its various functional clumps. It is true that vagal flow from the body shows a preference for the left insula over the right, and the left amygdala processes fear a little differently from the right. But it is enough for us to grasp the overall gist of the brain in a state of calm rather than the numerous qualifiers and details, which were, in my training, a major impediment to my own sense of calm.

The Happy Button

As Newton's Third Law of Motion states, all action has an equal and opposite reaction. And this applies nicely to the brain. Just as the brain has a sadness center, it also has parts that deal with happiness and pleasure and reward—even hedonism—including our Gatekeeper number 4, the nucleus accumbens. The power of hedonism is evident in several now classic studies on rats. More than fifty years ago, researchers Peter Milner and James Olds drilled holes in rats' skulls and implanted electrodes into their brains' nucleus accumbens (secured with jeweler's screws). The rats were trained to press a lever to deliver pleasurable sensation to themselves. One rat stimulated itself 7,500 times in twelve hours, almost twice per second, to give itself a dose of pleasure. These rats chose pleasurable stimulation over food and drink, eventually starving themselves to death in the pursuit of happiness. Although they lost out on life and liberty, they died happy.

Just as we need stress to survive, we need happiness to

succeed. The reward system, which also includes the amygdala, speaks to a wonderful thought: Happiness is evolutionarily built into our brain. Activation of these crucial regions of the brain is important in the pursuit of both calm and happiness. When sadness enshrouds us, the amygdala is overactive, as is the sadness center. The nucleus accumbens and its reward and motivation functions work in tandem with the anxiety-driven motivations of the amygdala to create a sea of calm, allowing us to sail to success.

The concepts of motivation and reward are crucial to our understanding of ourselves as sentient and feeling human beings. Are we inherently meant to be anxious? On face value, it would seem not. After all, we are more likely to remember happy memories and more inclined to forget unhappy ones, unless we are depressed, in which case we focus on the unhappy memories. However, the amygdala, the fear center, occupies such an elevated place in the hierarchy of our brain that I cannot help wondering if anxiety is a favored state for survival.

Did evolution design us to be anxious and unhappy most of the time? Is it in our nature?

I could think of no better person to pose this question to than Joseph Ledoux, a neuroscientist at New York University, who knows more about the amygdala than anybody else on the planet.

Dr. Ledoux replied with his characteristic erudition: "That may be the case in the short run," he said. "But ultimately, I think [evolutionary] pressure will drive the establishment of connections from lateral PFC [prefrontal cortex, frontal lobe] to the emotion regions [the gatekeepers]."

And then our emotions will be more controlled. But will that make us emotionless? Like *Star Trek*'s Mr. Spock? Is our destiny to become Vulcans?

He continued, "This could create 'Spock' type creatures—not the best adaptive route—or creatures who have a harmonious integration of cognition and emotion. Which is my preference. But," Dr. Ledoux adds wryly, "I'm not in charge of evolution."

Over time, it seems likely that evolution will increase the control our rational frontal lobes have over our unpredictable, emotional gatekeepers and core brain. It is a remarkable insight modern neuroscience has uncovered. But for now, we need to take these gatekeepers into account as we try to build a calm brain and a happy successful life. After all, while none of us is in charge of evolution, we are in charge of, say, how much sleep we get.

CHAPTER 7

The Mother of All Calm

"Methought I heard a voice cry 'sleep no more!
Macbeth does murder sleep'—the innocent
sleep,
Sleep that knits up the raveled sleeve of care,
The death of each day's life, sore labor's bath,
Balm of hurt minds, great nature's second
course,
Chief nourisher in life's feast."

—William Shakespeare, *Macbeth:* Act 2, scene 2

Today, sleep disorders and outright sleep deprivation plague us more than at any other time in history, leaving us with "hurt minds" and wounded bodies. Shakespeare knew something about the value of sleep, but science seems to have lagged behind the bard. In fact, the study of sleep is a relatively young discipline—until the mid-1950s, scientists believed that our brains lie dormant when we are asleep, that we are capable of brain activity only when awake. But thanks to an explosion of investigations since then, we now know a great deal more about sleep: why our bodies and brains crave it, why it's essential for both our equanimity and productivity, and why it can spell the difference between calm and chaos.

Our story begins, as so many stories in science do, with a few hapless rats. Back in the late 1980s, in the Windy City, researchers at the University of Chicago wanted to find out what would happen when an animal was deprived of sleep. They chose ten young male rats, attached electrodes to their heads to monitor brain waves, and put each one in a cage on a disc placed over a pool of water. As soon as the unfortunate rodents began to doze off, as revealed by the brain-wave monitoring, the disc started to spin, forcing the rat to wake up and move about so as not to fall into the water. Once the rat was fully awake (and the researchers knew this because the brain waves said so), the disc stopped spinning, leaving the bleary-eyed creatures safe from a dunking, but more tired than ever.

The rats' quarters were comfortable, they were able to exercise, and they had free and full access to water and food. But their lack of sleep had disastrous consequences. As the days went by, the physical changes in the rats were ominous. Their fur changed color from a "creamy white to brownish yellow . . . and stuck together in clumps as if it were oily"; they "looked disheveled," and patches of skin with ulcers were seen between the clumps of fur on the tail and paws. Their paws swelled and they had trouble walking.

In contrast, a control group of rats, which went without food for at least two weeks but were not subjected to the sleep deprivation, had "smooth, normally colored fur" and no skin ulcers, looking for all the world like "healthy, younger rats." While both sleep- and food-deprived rats lost weight, six food-deprived rats with lower weights survived longer than their sleep-deprived cousins. Clearly, weight loss alone did not account for their death.

Even as the sleep-deprived rats drank ever more thirstily and gorged themselves on the plentiful food, their weights plummeted. Their bodies were burning up nutrients faster than they could eat. The sympathetic system part of their stress response system went into hyperdrive to combat the stress of sleep deprivation, but still their stress response began to fail rapidly.

By the end of the second week of total sleep deprivation, two of the rats were dead, and by the end of the month, all of the sleepless rats had expired. During the last twenty-four hours of their sorry lives, they lost the ability to regulate their body temperature. By the time of their deaths, they had succumbed to the invasion of their body and blood by erstwhile friendly bacteria from their gut. Their stress response systems threw in the towel, unable to withstand the sleep deprivation, even though there was adequate provision of all other nutrients and comfort.

This sad tale speaks to how crucial sleep is for rats, more important even than food. But, of course, the goal of this study and others like it was to draw inferences about the effects of sleep deprivation on us humans. With sleep disturbances reaching epidemic proportions, researchers are particularly interested in studying sleep, which, despite all that we know about it, still remains elusive, the mysterious princess of the night.

Before the piteous sleepless rat experiments, back in 1958 in the City of Angels, Los Angeles, researchers offered financial inducements to volunteers to stay awake as long as possible. Four young men, ages twenty-one to twenty-three, signed on and checked into a brightly lit, large research ward for observation, where they were watched constantly by nurses. The unit was equipped with a Ping-Pong table, television, card games, and snacks.

After being given a battery of tests, including measures of their memory, they were individually monitored over the ensuing days; electrodes pasted to their scalps allowed their brain waves to be checked during specific tests. The young men were observed by the research staff over the entire time of the experiment, except when they used the toilet or shower, and their social interaction was recorded. Their memory was tested periodically. A game to test group interaction and communication patterns was played by the volunteers at 50, 100, 150, and 200 hours of sleep deprivation. A complete personality battery of questions was asked of the subjects at the beginning and at 200 hours of sleep deprivation. Daily tests of orientation and cognition were conducted. Hand-eye coordination was assessed with a test requiring the subject to maintain a wandering two-millimeter spot of light within a ten-millimeter circle, using a handheld control.

By the second day, there was an increase in fatigue and disruptive periods. By the third day, reading became impossible because of "the subjects' inability to maintain concentration." Also, starting on the third day, all four men reported blurry vision, which took the form of "smoke or haze rising from objects."

By the fifth day, they seemed to get a "second wind." They were able to cope better with staying with the experiment but continued to have increasing numbers of "lapses"—including misperceptions, hallucinations, deterioration in cognitive performance, moodiness, and even occasional "regressive behavior." They started to complain bitterly about the food, put up protest charts around the ward, and call the medical student assigned to draw their blood Bela Lugosi. They began to "mess" with food as children would, picked fights with one another, and required increasing amounts of noise and light to stay awake.

However, between these lapses, the young men were able to pull themselves together enough to perform adequately the various cognitive and motor tasks asked of them. By 200 hours of sleep deprivation, the personality test revealed them to have higher scores for psychosis and schizophrenic patterns. They floundered on tests of logic.

They developed a tremor and jerky eye movements. The hardest time to stay awake was between two A.M. and six A.M., and they devised a method of combating sleep by immersing their face in a basin of ice cubes. By day three, the eyelids began to droop, and by day seven, speech began to slow. Physical activity was slow and was interspersed with periods of "frenetic" activity. Their lapses coincided with an increase in slow brain waves, usually seen in deep sleep.

Then, at seven days ("168 hours"), as he was performing the hand-eye coordination test, one young man "screamed in terror, pulled his electrodes off, and fell to the floor sobbing and muttering incoherently about a gorilla." Upon questioning, he was found to be reexperiencing a recurrent nightmare he had as a child of five, in which Humpty Dumpty was menaced by a gorilla. He explained later that, as he was tracking the point of light on the screen, it had morphed into Humpty Dumpty and then morphed into the gorilla, which then began to threaten him. Before he tore off his electrodes, his brain waves showed activity consistent with wakefulness. Sleep deprivation had led this young man to literally live out his nightmare.

By 200 hours, or after more than eight days, the group seemed to be "nearing the end of their endurance," and the researchers decided to terminate the experiment. The young men were followed for some time after the conclusion of the experiment and did not suffer any long-term ill effects.

An earlier study of medical students who were sleep

deprived for seventy-two hours, found them to be more irritable, with visual hallucinations and slowed cognition. While these were all extreme situations that few of us will ever experience, they speak to the importance of sleep for clearer thinking and a better mood. It is hard to be calm when running on less sleep.

The Less-Sleep Habit

Extreme sleep deprivation, like that imposed on our young men and young rats, can lead to serious cognitive, psychological, and physical impairment. Even though the young men could not be kept awake indefinitely, as was done with the rats, it is clear from the research that maintaining a state of constant wakefulness is impossible. However—and this is crucial for our purposes here—even moderate sleep loss, especially if it is chronic and persistent, can produce an alarming and debilitating set of symptoms, from anxiety, irritability, and depression, to impaired decision making and problem solving.

Why is sufficient sleep essential for maintaining a state of calm? The answer lies in both the brain and the body. During sleep, your body is in a state of complete relaxation, but your brain continues to be active. Although the activity is regimented and synchronized, with vast portions of the brain all humming in unison to the same conductor, parts of the rational frontal lobes that deal with judgment and logic are switched off. Vagal tone is also high during sleep as breathing and the heart rate slows. To promote this state of complete relaxation, a switch gets turned off in the core brain, and the body becomes *paralyzed.* Sleep loss interferes with these processes and makes it difficult to respond to emergencies, to pay attention, and to assess rapidly changing conditions, as when a trucker pulls an all-nighter or a surgeon works on little sleep.

Sleep loss causes mental inflexibility and rigidity, trapping you in a rut and preventing you from thinking your way out. Given what happened to the sleep-deprived rats, it is also very likely that sleep helps with the human immune system and with healing. Because the stress response is mounted in reaction not only to loss of sleep but also to, say, the stress of pneumonia, it is easy to see why chronic sleep deprivation is likely to leave you less able to fight off infections.

Chronic sleep loss often begins in adolescence or during freshman year of college, when young people are away from home for the first time. It then develops into a habit that may be difficult to break.

A peppy, sunny girl named Lucy lived on my block. An only child and a straight-A student, Lucy maintained a strict schedule, under her parents' watchful eye, of school, swim meets, volunteering at the local senior center, and home-cooked dinners. Even during her high school years, Lucy's adoring mother, Mary, helped her maintain a balanced routine, with some time for socializing but enough time for relaxation. Of course, this was all before Lucy went off to college.

Two months into her first semester, her parents told me the phone rang at one A.M. one night.

"Mom. Oh, Mom!" Lucy was on the other end of the line sobbing.

"Lucy! What's wrong?" Alarmed, Mary nudged her husband awake. "Is everything okay, honey?" Lucy's father called into the phone. "Are you hurt?"

"No, I'm fine. I mean, I'm not fine. But I'm not hurt or anything like that. I just can't sleep! I haven't slept in three days straight. I have an economics midterm tomorrow and I need to sleep and I can't! I am soooo tired," she wailed.

After Mary and her husband got over their initial fears and were reassured that Lucy was all right, they asked her more questions. It turned out that she had taken to staying up late with her dorm mates, eating and talking and laughing into the night. She had gotten into the habit of staying up sometimes until two or three in the morning. Always a conscientious child, as her midterms neared, she became more and more anxious, and when she tried to get to bed at nine P.M., sleep did not come. Her recent bad habit had taken hold. Then she began to worry about her exams, and this made matters even worse. Finally, in desperation, Lucy called her mother.

"Erica down the hall has a lot of Ambien, Mom. She said she'd give me some. I know you never liked me to take drugs, Mom," Lucy pleaded, "but I can't take this anymore!"

Lucy's situation illustrates a common type of sleep disorder that is generally caused by a shift in sleeping habits, accompanied by anxiety or concern about a stressor. It not only reduced Lucy, a normally sensible and independent young woman, into a sobbing girl calling her parents in the middle of the night, it made her more irritable, moody, and impulsive.

It is no surprise that sleep and calm are inextricably connected. Consider:

- Over 90 percent of depressed persons complain of some type of sleep disturbance.
- Insomniacs may be up to ten times more likely to be depressed and up to seventeen times more anxious than people who sleep well.
- Interestingly, sleep deprivation is associated with obesity. Certain nerve cells in your brain's hypothalamus get excited when you are stressed. This makes you stop sleeping and overeat at the same time.

Many people think of sleep as wasted time, time that could be used to do other, more "productive" activities and to interact with friends and family. One of the core causes of the widespread plague of sleep disorders is this devaluation of sleep. Need to finish that paper? Let me cut down on my sleep. Need to hang out with my friends and still get to work on time? No problem, I can just sleep less. Need to watch my favorite late-night television show and still be up in time to make breakfast for the kids? Hey, I can always cut back on my sleep. Unfortunately, this type of thinking deprioritizes sleep, eventually leading to chronic insomnia and, as we have seen, a score of other problems.

Few activities are as productive and interactive as sleep. It is a time when your brain interacts with itself, fine-tunes what it has learned, plays back daytime experiences to help you remember them, and lives out dreams—and you awaken renewed, refreshed. A good analogy is the concept of debt. Debt is debt, whether it is tangible money or credit card debt. I think many people view sleep in the same way they may view their credit cards—as an unlimited source of time (money) that they can draw on without repercussions. But just like real debt, built-up sleep debt comes back to haunt you, causing a pervasive lack of calm—which, as we know, leads to a host of other problems.

Despite all we know about what happens when we don't sleep enough, and despite all the research into sleep that's been conducted in the last sixty years, the core functions of sleep still elude us. Did evolution design sleep to protect us from nighttime predators? To allow us to perform necessary brain and body maintenance? To enhance our ability to think and to remember? All of these are true, but none alone tells the full story. What we do know beyond any doubt is that, like so many other essential functions, including breathing and the stress response, sleep is an adaptive state that has been conserved across species and over millennia. It is essential for our survival. Cockroaches go to sleep, fruit flies go to sleep, and worms do, too. In fact, no animal has yet been described that does not need sleep, and all animals, when deprived of sleep, need recovery sleep. Recovery sleep, which is the extra sleep you might need after spending a couple of nights out on the town, say, is basically "catchup" sleep.

Sleep is a time of adaptive inactivity when the body allows us to be more efficient and use energy wisely. This energy is intelligently conserved when we are most at risk of being attacked by a predator and have minimal vital needs—as in the dark of night. Children sleep more hours to conserve the energy required by their higher metabolic rate. By reducing muscle tone and effectively paralyzing the body, sleep leaves you limp and reduces the risk for predation and injury at night from acting out dreams. Had the young man who ripped off the electrodes and fell to the floor actually been asleep, the sleep-induced paralysis of his body would have kept him inert and safe in his bed.

Unlike hibernation, sleep lets you awaken quickly to deal with emergencies and exigencies—like a real gorilla or a crying infant. In ancient times, sleep was the state in which mothers nestled close to their offspring and kept them relatively immobile against nighttime predators. Darkness was also a time for socializing. In primitive societies, celebrations often took place at night, and sleep was a communal activity. People dropped off to sleep exhausted on the periphery of festivities in the village. But time and space were more fluid. One might return to work a few days later and stay asleep very close to home. Moreover, all the members of the tribe participated in the celebration, so there was no accounting and no guilt.

In recent times, late-night communal activity is common, but returning to the dictates of the alarm clock and to the pressures of a world unsympathetic to one's nightly excesses prevents you from catching up on sleep. Lucy had to take her examination; her professor would not have understood that she'd had a pizza party the night before. And few bosses look kindly on staff wandering into the office at ten A.M. Or even nine A.M. Nowadays, the traditional nine-to-five often turns into the nontraditional but expected eight-to-six. A long workday can leave you tense and edgy and not calm enough to fall asleep easily.

This is the disconnect between modern life and ancient life. Celebration in days of yore was not followed by a sudden and rapid return to work but a more gradual awakening and folding into daily activities. Sleep was not sacrificed and there was no attendant anxiety about having to be somewhere even when the body and brain would rather be where they belonged that morning—in bed. When there are no punitive consequences to a night of celebration—physical, social, or mental—calm reigns.

The amount of time one sleeps is largely inherited and the day-to-day variations in sleep are generally small. Daily naps also influence how much sleep we need at night.

Because of his afternoon naps, Winston Churchill was able to stay up till the early hours of the morning. These days, few of us can avail ourselves of this luxury, but even a short nap during the day can be restorative.

Contrary to the conventional wisdom, the amount of time you sleep is not affected by your level of physical or mental activity. Whether you toil through today's typical sixty-hour workweek or simply hang out all day at home, you will sleep about the same amount each night. The only constant predictor of increased sleep is previous sleep deprivation. So if you've been pacing the floor on Sunday night, worrying about the report due the next day, you will probably want to turn in early on Monday. Will your life let you?

The Goodness of Sleep

Just like the amount of time you sleep, all your other sleep traits, including the amount of time you dream and the length of your sleep cycle, are inherited and, interestingly, inherited independently. As with all inherited traits, there is greater similarity in sleep between identical twins, who share the same genes, than between fraternal twins, who are no more alike genetically than ordinary siblings. What this means is that the amount of time slept is inherited separately from how much time you spend dreaming, just as eye color is inherited separately from the length of your eyelashes.

Sleep is essential for laying down new memories, practicing new skills, and correcting daily mistakes so that the future can be shaped by past experiences. At night, little finches silently play their birdsongs over and over again—polishing their warbles to achieve perfect harmony for the next day. Humans do something like this, too. Violinists practice the violin as they sleep, and dancers glide on ghostly feet. Surfer dudes ride the big wave out safely in the night, infinitesimally moving a muscle here and there to make the next day's surfing even more spectacular. Jack Nicklaus went through a rough patch of golf until he dreamed of holding his club a little differently. And voilà! Problem solved.

Some of us continue to play Tetris or trawl the Internet in our sleep, if that is mostly what we do when awake. Scientists know this because they have measured activity in different nerve cells while our finches are singing and while they're sleeping. They also perform experiments in sleep labs in which they rouse student volunteers who spent all day playing Tetris just as they start dreaming and find them playing Tetris in their sleep as well. Any one of you who reads avidly has surely had dreams in which you read books. Our brain is very active indeed during sleep.

For much of civilization, most people, with the exception, of course, of the prescient William Shakespeare, viewed sleep as an inactive state, almost a state of hibernation, the opposite of waking and activity. In the early 1950s, however, again in the Windy City, an important discovery revolutionized our view of sleep. Russian physiologist Nathaniel Kleitman and his graduate student Eugene Aserinsky discovered that sleep was not a monolithic whole—as was previously believed—but consisted of two distinct stages. Aserinsky had noticed that when people slept, their eyelids twitched sometimes and not at other times. This eyelid movement was correlated with brain-wave activity and revealed two distinct types of brain-wave patterns. Kleitman and Aserinsky named these stages rapid eye movement (REM) sleep and slow-wave sleep.

By observing and monitoring volunteers in a sleep laboratory and by waking them up at various intervals, the scientists learned that people sleep through several consecutive cycles of these sleep stages, on average four a night. Because brain activity during REM sleep so resembles that of the waking state, even as the person is sound asleep, REM sleep is also called paradoxical sleep. In the irony that is real life, Eugene Aserinsky, widely hailed as one of the fathers of sleep research, was killed when his car ran into a tree: Aserinsky had fallen asleep at the wheel.

The importance of the sleep stages is only now being slowly understood. It is during slow-wave sleep that your brain records memories of facts and episodes, like a movie you saw today or what you learned in class. During this sleep stage, basal metabolic rate and body temperature drop, and the discharge from the vagus nerve is high. When the vagus is stimulated, the heart slows down and is less responsive to the effects of stress hormones. The bowel is relaxed, the diaphragm and lungs function more slowly and deeply, and the airway passages open up to let more air in. The vagus nerve also controls most of the muscles of the throat.

In many ways, then, slow-wave sleep, which is naturally wired for slow, deep, rhythmic breathing and quiet, powerful heartbeat, is the state that meditation replicates externally. In other words, during effective meditation, vagal tone overrides the hormones associated with the fight-or-flight response and other stress hormones.

However, deep slow-wave sleep is strikingly different from meditation in one key respect. In a series of interesting studies, Richard Davidson, a psychologist at the University of Wisconsin, demonstrated that experienced practitioners of some Tibetan traditions of meditation had rhythmic oscillations of brain activity, a state similar to sleep. However, the dorsolateral prefrontal cortex, an area of the brain that is important for judgment and higher cognitive function, is turned off during sleep and is activated during experienced mediation. This means that meditation, while causing the brain to hum as one, much as sleep does, is not a state of sedation but a state of heightened, focused awareness. Sound familiar? Focused awareness is a perfect description of "relaxed and confident alertness," namely, calm. This may be one reason that "monks don't frown."

During REM sleep, in which your eyes move rapidly back and forth, the basal metabolic rate and body temperature rise again, and some areas of your brain become more active than when you are awake! Your body then turns off the core-brain switch that is responsible for muscle movement, effectively paralyzing all muscles except those controlling your eyes, your breathing (the diaphragm), and the muscles that move your inner ear bones—presumably so that you can be warned if there is any danger nearby.

There are other ways in which REM sleep is different from the waking state. During REM sleep, your breathing is unresponsive to rising blood carbon dioxide levels. In contrast, when you are awake, your breathing becomes more rapid, to blow off the carbon dioxide. Also, during REM, we become cold-blooded animals, losing our ability to regulate our body temperature and drifting toward ambient, external temperatures. (This is what happened to the sleep-deprived rats—they couldn't control their body temperature when they were *awake.*) In other words, your body becomes metaphorically unhooked from your brain and floats off into the wider world. This is why marine mammals, such as dolphins, alternate REM from one hemisphere of the brain to the other, so that they can continue breathing and stay alive while sleeping.

Why does your body disengage from your brain during REM sleep? The obvious answer is so that you don't act out your dreams while you sleep, as did our young man who was awake when he dreamed of the gorilla. Else, we would be up and running at night, risking injury, as we chase the unicorn that beckons us into the forest. A fascinating fact is that dreams remain real to us, something that we have no learning curve for; we never learn how to tell that our dreams are just dreams. From an evolutionary perspective, it is noteworthy that throughout the millennia, the brain never modified itself to be able to distinguish between imaginary dreams and reality as we sleep. Could it be that perhaps we would breathe faster when being chased in our dreams and this would disrupt carbon dioxide levels in the blood? If so, it may be better to disconnect the body from the brain and have it switch itself back on in an hour or so, when we return to slow-wave sleep.

Interestingly, though, and contrary to what most of us have been taught, dreams occur during both REM and slow-wave sleep. Non-REM dreams are shorter, less vivid, less emotional, and more coherent than REM dreams. However, narrative coherence itself is not a hallmark of most dreams, in either REM or slow-wave sleep. As one observer said, "Dreams are not successive chapters in a book but are rather separate short stories." Another curious thing about dreams is that while most are in color, 20–30 percent are monochromatic. I do not know if this has any bearing on calm, but one could speculate that colors of the same shade tend to be more soothing than the full rainbow.

Sleep and dreaming are important aids to learning. The optimal way for students to do well on exams would be to study every day and then sleep for a minimum of eight hours with no interruptions. If you are the cramming sort, go ahead, but make sure to sandwich a nice, long sleep session between the cram and the test. This type of sleep helps consolidate all those bytes of information and occurs best with so-called natural sleep. What this means is that sleep induced by hypnotics and prescription medications does not replicate the architecture of natural sleep, which is the optimal state for memory consolidation and learning. Natural sleep and natural awakening do not leave you with a hangover or jangling nerves; you awaken at precisely the time your body is ready to get up and go. There is a sense of complete rest. Natural sleep, as opposed to drug-induced sleep, does not interfere with the complex mix of neurochemicals that are secreted and absorbed during sleep.

Pills, Pills, Pills

"That's all very well," I can hear some of you say indignantly. "But what if I can't sleep at all? Isn't it worse for my sense of calm if I don't sleep at all than if I get some sleep with Ambien?" The short answer is yes, of course; nothing is worse than total sleep deprivation. However, relying on drugs to sleep is a slippery slope, leading to more drugs for less sleep, as we saw with Lisa. I advocate anything that increases vagal tone, including hot baths, a massage, meditation, as well as aids like melatonin that do not interfere as much with the sleep cycle. Also very helpful is keeping your television and other electronic devices out of your bedroom, including cell phones. Leave them in another room and change the ringtone to something less alarming—music or the more familiar ring of a landline. Beeps jar the core brain into high alert, making it difficult to fall back to sleep. Personally, I have found the steady hum of a ceiling fan to be an excellent sleep aid and others swear by "white noise" machines. The sound frequencies they generate, similar to the humming and chanting in several meditative and religious practices, may be soothing and calming to the core brain.

An important caveat to bear in mind is that we all have an individual amount of required sleep, a "private" sleep cycle, with some people requiring eight or more hours and others a mere six. This does not mean that those who sleep more are "wasting" away that extra time. It is easiest to be productive and efficient with the *right* amount of sleep for your brain, not the *least* amount of sleep. I, for example, require about nine hours of sleep most nights for optimal efficiency. When I sleep less, I waste time during the day doing shoddy work that takes twice as long as it would have had I slept as much I needed to.

Sleeping and dreaming can also enhance imagination and innovation. Want to be a creative thinker? First, work on the problem and immerse yourself in the many facets of it. Then fall asleep. Nineteenth-century chemist Friedrich Kekule accidentally and famously discovered the phenomenon of coming up with ideas while you sleep. Kekule and his fellow chemists wrestled unsuccessfully with the structure of benzene, until Kekule fell asleep and dreamed of a snake eating its own tail. He then realized that the benzene molecule was ring-shaped. For Kekule and countless other creative thinkers, "sleeping on it" is much more than a metaphor.

For some people, sleep gets sidelined because they are "too busy." Moms with young children, managers with high-pressure responsibilities, physicians in the emergency room, all have reasons why they could be awake instead of sleeping. But sleep is a time for creativity, for "lateral thinking" in which you look at problems from a novel perspective. It is the time for consolidation of facts and the weaving of those facts into a vibrant tapestry, unrestrained by logic, that is the fount of creativity.

In spite of the overwhelming evidence that sleep is essential to our well-being, productivity, and creativity, indeed, to our survival, a surprising number of people—twice as many women as men—can't or don't make time for downtime. In 2010, some sixty million Americans were sleep deprived. Interrupted sleep—which causes disruption of the sleep cycle and prevents true, deep sleep—is rampant. In our anxiety-ridden culture, more and more people fail to fall asleep, or stay asleep night after night, not realizing that their cognitive functioning and their mood are below par simply because they are not properly recharging their mental engines. Many people even dismiss sleep as akin to sloth, boasting about how little rest they "need" and claiming their careers would be jeopardized if they turned in early. We all know workaholic insomniacs who, unable or unwilling to take to their beds, dispatch e-mails at three A.M.

Is a good eight to nine hours of sleep a night likely to cost you your job? Winston Churchill weighed in on this question, declaring, "Don't think you will be doing less work because you sleep . . . that's a foolish notion . . . you will be able to accomplish more." Even Thomas Edison, who publicly disdained long periods of sleep and was devoted to short daytime naps, was, to surmise from his diary jottings, a restful nighttime sleep enthusiast. Our cultural denigration of sleep is seen in the conflict the Dalai Lama has during his visits to the West. He is known to sleep from nine P.M. to three A.M. and then meditate from three A.M. to seven A.M. However, because he is constantly invited out to dinners in the United States, he finds it difficult to maintain this schedule.

Sadly, for adolescents and young adults, there is a cultural stigma about going to bed at eight or nine P.M. Aside from the pressure of schoolwork, socializing often begins at this time or later, and those who opt to sleep may lose out. Confronted with this dilemma, most students sacrifice sleep. Furthermore, adolescents may be more susceptible to the mood-altering effects of chronic sleep deprivation. Those young people who get less than the required nine hours of sleep—yes, nine hours for young adults—are more prone to depression and suicidal thoughts than their well-rested peers. Interestingly (and reassuringly), surveys reveal that adolescents generally conform to curfew times set by their parents. Good sleep habits may therefore be molded by parents, helping their children reap a lifetime of benefits.

It is becoming more common for adolescents and college students to resort to medication to sleep, as Lucy suggested to her mother. Sleeping aids are one of the most common requests for medications I receive. Internists and family practitioners are even more inundated with requests for sleeping pills, and prescriptions for sleep aids and anti-anxiety medications are sharply on the rise.

A *Los Angeles Times* article from March 2009, entitled "Sleeping Pill Use Grows as Economy Keeps People Up at Night," lists lost jobs, failing businesses, and college acceptance dilemmas as culprits that keep Americans awake. In fact, according to the National Sleep Foundation, over a quarter of Americans blame anxiety and stress about the economy and personal finances for keeping them awake.

In the last decade, there has been a 50 percent increase in the use of sleeping pills in adults under forty-five, and particularly college students. This group is fast replacing the elderly as the population with the steepest increase in sleeping pill use. Not surprisingly, anti-anxiety–type drugs, which are used as sleep aids, constitute the most common type of prescription filled in the United States. We are more sleep deprived and less calm than ever before, at least gauging from our use of prescription drugs.

The number and variety of sleep aids and anti-anxiety medications have also proliferated. Here's a little quiz for you. Recognize any of these names?

Diazepam
Clonazepam
Lorazepam
Alprazolam
Temazepam

I thought not. How about these?

Valium aka diazepam
Klonopin aka clonazepam
Ativan aka lorazepam
Xanax aka alprazolam
Restoril aka temazepam

I thought so. These brand names are so familiar, so much a part of our national consciousness, that most Americans are able to identify them in their sleep. Pun intended!

The umbrella term for these drugs is *benzodiazepines*. Here's what happens to your brain on "benzos," as they are usually called. Ready?

Your brain is a maelstrom of busy nerve cells chattering away, using chemicals that excite or inhibit each other. The mother of all inhibitory chemicals in the brain is GABA (Gamma AminoButyric Acid, for those of you who want to know), and benzos work by upping GABA's effect. So, in theory, a little benzo makes you mellow, a little more makes you sleepy, a little bit more makes you pass out. You get the idea. Reality, however, is another story.

1. Benzos cut down on your REM sleep. That's when you rehearse your daily activities, like how to tie your shoelaces if you are a toddler, how to drive with a stick shift if you are a teenager, or how to hit the ball out of the park if you are Babe Ruth. Being on benzos means you will most likely strike out.

2. Benzos reduce slow-wave sleep, which is the deepest sleep. Since this kind of sleep is important for the consolidation of facts and events, being on benzos means you are less likely to ace that exam or make the right calls on the trading floor.

3. The more you use benzos, the more you will use benzos. You will need larger doses of them and more often. You may even get to the point (and many people do) where you cannot sleep without them.

4. Benzos exacerbate sleep disorders like sleep apnea, a condition in which a person stops breathing momentarily while sleeping, often

hundreds of times each night. This causes long-term health problems, including depression, anxiety, and hypertension, not to mention, of course, a poor night's sleep. And all the increased snoring from sleep apnea will bode poorly for your spouse or partner who may then need to go on benzos themselves.

5. Being on benzos long-term can produce memory loss that mimics dementia.

6. Chronic benzo use causes daytime sleepiness, reduced concentration, irritability, and anxiety. These symptoms occur regardless of dosage.

The sad and simple truth is that many of the chronic effects of taking benzodiazepines for sleeplessness sound a lot like the effects of chronic . . . sleeplessness. As much as possible, avoid using these pills.

Sleep is a state of concerted brain activity during which several crucial processes necessary for optimal daily functioning occur. It is the mother of all calm.

CHAPTER 8

The Stress of the Two XX's

My vagal tone was climbing higher and higher as I listened to a lecture on "physicians' quality of life" at the annual meeting of the American Medical Women's Association. The genial tone of the speaker, the warmth of the room, and the proximity of my colleagues all conspired to produce in me a pleasantly drowsy state that was in danger of heading further south. Suddenly, I was jolted awake by my barely vigilant frontal lobes, which had just processed what the speaker was saying: "What! How could that be?" I said to myself. Women physicians in the United States are 250 to 400 percent more likely to commit suicide than other women? My core brain and amygdala reacted viscerally to this unwelcome piece of news, my frontal lobes turned on their skeptic alert system. My vagal tone plummeted as I sat up, fully alert, and gazed around the room. The female physicians in the audience looked in a chipper mood. But appearances can be deceiving. Were these women really more at risk for suicide?

To put this blunt statistic in perspective, the speaker added that male physicians were also more likely to take their own lives than their nonphysician peers, but at a rate closer to 70 percent, far lower than their female counterparts. True, practicing medicine is no walk in the park, but the rough road is not just a "woman's issue." Both male and female physicians have to cope with the daily exposure to pain and suffering, to the threat of lawsuits, to the countless other pressures of a high-stakes profession. So what would account for the shocking number of female suicides? I knew that nearly half of women doctors report being sexually harassed (the number is a staggering 75 percent for surgeons). More female than male physicians are single and childless, both risk factors for suicide. Even so, it was hard to fathom what was driving my sister physicians to kill themselves.

As the speaker pointed to the harsh statistics on the PowerPoint screen, I began to think about women outside the realm of my own profession. The consensus among experts is that women are anywhere from two to six times more at risk for anxiety disorders than men. As early as the age of six, future women doctors were already twice as likely to be anxious as their fellow boy first graders who were also destined for a career in medicine. What is it that makes girls and women less likely to achieve calm and equanimity throughout life?

I knew that men are more susceptible to high blood pressure, heart disease, and drug and alcohol abuse. Women are more vulnerable to what I call the slow-burn diseases: autoimmune problems, chronic pain, anxiety, and depression. Brain neurochemistry may in fact predispose women to emotional disorders. Low levels of the brain chemical serotonin are seen in patients with both anxiety and depression, and serotonin production is higher in men. Antidepressants like Prozac act by increasing serotonin function, and 16 percent of American women use these drugs, nearly three times more than men.

Gender differences in the serotonin response to stress has been studied. In one experiment, researchers briefly removed a litter of degu pups from their mother. For these South American rodents, which live in large social colonies, even temporary separation can be upsetting. The researchers allowed the pups to hear their mothers' call while they were separated from her and found that this increased the serotonin receptor concentration in the male pups' amygdala but decreased the concentration of these same receptors in female pups.

Although we can't always extrapolate from studies like this to human behavior, these results suggest that boy and girl babies might also react differently when they are temporarily separated from their mothers. Girl babies may be less resistant to separation anxiety, which can set the stage for anxiety disorders later in life. The sensitivity of women to stress and attendant increase in anxiety is also linked to estrogen, which is "switched on" at puberty and diminishes after menopause. Estrogen influences the activity of brain chemicals like serotonin, resulting in the mood changes that some women and girls experience during or just before their periods.

Biology doesn't tell the whole story, of course. Girls and women are also at greater risk for sex abuse and more sensitive to the effects of adverse childhood experiences. Moreover, they must grapple with increasingly complex and conflicting social roles. In fact, among homogenous groups like the Orthodox Jews of London or the Old Amish in Pennsylvania—where gender roles are clearly determined—there is no difference in depression and anxiety between men and women.

Beyond these cultural and hormonal factors, there may also be neuroscience-based explanations for gender differences in achieving calm. Does the particular way women's brains process stress make them more vulnerable to anxiety and lack of calm? Is there a difference between men and women in cortisol secretion, and the recruitment of the parasympathetic and sympathetic systems that explains why my female colleagues are at higher risk for suicide? Does vagal tone differ between men and women? And are there differences in the response of the amygdala and insula? The answer to all these questions is yes—and when we take a closer look at each of these factors, we see that the differences can sometimes be dramatic. Consider the story of Helen, a cheerful, hardworking woman who was lauded as an expert seamstress at our neighborhood dry cleaner's.

Helen is the fifty-year-old single mother of Elena, an adorably irrepressible teenage minx of "sixteen going on twenty-eight," as Helen described her. One night, while she was working the late shift at her second job as a cashier at her local supermarket, Helen received a phone call that changed her life. Elena had been in a car accident and was in critical condition at the hospital.

In a daze, Helen slowly removed her apron and walked out to the parking lot. Elena's crazy gap-toothed smile, the way she licked her ice-cream cone, their latest fight about an outfit that Helen considered inappropriate, all these images flashed before Helen's eyes.

"My baby. My baby. Oh no! Not to my baby," Elena kept repeating as her coworkers helped her walk to her van. She was almost there when she collapsed to the ground, clutching her chest with pain. An ambulance rushed Helen to the same emergency room that her daughter had been taken to an hour earlier. In the hospital, doctors diagnosed Helen with heart failure and a heart attack. It was only later that they discovered that Helen had been struck down by Takotsubo's cardiomyopathy, or broken heart syndrome.

This little-known condition was described for the first time in Japan in 1993. What is unusual about it is that 90 percent of those afflicted are women, most frequently women over the age of fifty. In the United States, nearly 6 percent of women diagnosed with heart attacks may have this condition. There is reduced vagal tone in postmenopausal women, the group that this condition strikes the most. It appears that these women have a particular cardiac susceptibility to stress (emotional or other) and that this causes damage to muscles of the heart, leading to heart failure. The good news is that most patients recover completely within two months, as Helen did, back to mending clothes with her old smile in place. The irony, though, is that women are far more likely than men to be misdiagnosed as having anxiety when they are actually suffering from heart disease.

How did anxiety and sudden stress cause heart failure in Helen? When Helen was at the cash register, working as usual, her heart was exposed to lower levels of sympathetic activity and she had lower levels of adrenaline than a man her age. Her vagal tone was higher than that of a man. But when she heard the news of her daughter's accident, her adrenaline level rose dramatically, "stunning" the heart and causing heart muscle damage. At resting state, men are predisposed to be more "vigilant" than women, so when stress occurs, their core brain and sympathetic nervous system engage, but the proportionate rise in adrenaline is not as high as it is in women, whose sympathetic nervous system activity is lower to start with. An additional risk may be the smaller size of the heart chambers in women.

So, although women have higher vagal tone, speaking to their evolution as mothers and keepers of peace, their response to stress is far more extreme, and, in the case of this particular cardiomyopathy, their heart muscles crumble under the flood of adrenaline. Men, soaked in a stronger adrenaline brew than women, are not as susceptible to the sudden effects of high levels of adrenaline. However, chronically high levels of sympathetic activity and low vagal tone cause more men to die from heart attacks than women. Even while they are sleeping, men have lower vagal tone than women, which may account for their early-morning cardiac deaths.

Vagal tone is also related to estrogen. The fact that broken heart syndrome strikes postmenopausal women more often than younger women or men of any age suggests that hormonal activity may be involved. Estrogen has been shown to reduce pathologic changes in heart rate in rats, and estrogen supplements can partially prevent the kind of stress-induced cardiovascular assault that Helen suffered.

The Cuddle Hormone

Further confirmation of the connection between vagal tone and estrogen is provided by a study in Taiwan of about one thousand men and women between the ages of forty and seventy-nine. The researchers found that women had more parasympathetic activity and a higher vagal tone, while men had more sympathetic regulation of the heartbeat. This higher vagal tone in women disappears after menopause but recurs with hormone replacement therapy. But estrogen isn't the only hormone, and there is perhaps a more important one when it comes to producing a calm state.

Women have traditionally benefited from bonding in groups for protection and support. In ancient times, this mutual care could mark the difference between life and death for women and their children while the men were out hunting or battling enemy tribes. The psychologist Shelley Taylor has coined the phrase "tend and befriend" to describe women's powerful coping style of interacting and supporting one another, in distinction to men's fight-or-flight reaction.

Taylor attributes women's tend-and-befriend pattern in part to the hormone oxytocin, released during childbirth and breast-feeding, during touch, and during orgasm. (Interestingly, even in men, orgasm stimulates oxytocin and promotes sleepiness and calm.) Breast-feeding mothers are more likely to describe positive mood than mothers who bottle-feed exclusively. And in studies in which a woman breast-feeds before being exposed to a psychological stressor, her stress response is lower than in nursing mothers who were also bottle-feeding.

Taylor's work has helped me understand why it's so important for girls and women to take comfort in sisterhood and why it is essential for people like me to be in the community of other physicians at a place like the American Medical Women's Association annual meeting. I have come to believe that women need friends more than men do in order to keep calm.

An abundance of research speaks to this need. In one study, prairie voles—interesting creatures committed to monogamy—were treated with oxytocin before being placed in a situation that mimicked a flooded burrow. Then they were removed to a dry cage to recover, either by themselves or accompanied by a "friend." Even with a preemptive dose of the tend-and-befriend calming hormone oxytocin, the voles that recovered alone appeared visibly anxious, trying frantically to escape. Both male and female voles who recovered with a friend got better faster, and their oxytocin levels were higher as well. In women, the balm of oxytocin works best when mixed with the ointment of companionship.

Sometimes known as the "cuddle hormone," oxytocin increases sociability and bonding not only in women but also in men and promotes calm through a combination of increased vagal tone and reduced critical thinking and fear. Oxytocin also boosts trust even when that trust is misplaced. In a study in which women and men were given nasal sprays of oxytocin, they continued to trust strangers with their money even after these strangers were revealed to have mishandled their funds. Brain imaging showed that oxytocin reduced activity in the brain's fear center, the amygdala, and in brain areas involved in decision-making.

Oxytocin increases generosity and empathy, allowing us to look at a situation from another person's perspective, in the process reducing anxiety. It also helps us to read the mental state of others using cues from facial expressions. This should not be surprising since we have oxytocin receptors in the brain's "gut feeling" center, the insula, which is the primary endpoint of most things vagal.

Even as oxytocin makes both men and women more generous and, for example, increases the amount of money returned to a person we know who helped with an investment, it does not increase the amount of money given when the return investment is determined by random lottery. It seems that oxytocin preferentially promotes more bonding behavior in those we know than in those who are strangers.

Oxytocin promotes relaxation and supports coping skills. In rats who were forced to swim in a stress test, oxytocin helped them be calm enough to float and stay above water longer. Oxytocin may also be associated with the ability to endure trauma. In a study of twenty-two women aged eighteen to forty-five who had endured childhood abuse, the greater the degree of abuse, the lower the level of oxytocin in these adult women.

In conditions where there's a lack of calmness and an inability to tolerate change, such as autism, oxytocin may be helpful not just in increasing calm but also in helping people with autism to interact socially. Social interaction is not limited to human connections. I have brought my dogs to my medical office for many years, holding the firm belief, bolstered by research as well as my own observations, that dogs make patients feel more comfortable and calm, reducing the typical nervousness that accompanies a visit to a physician. One study found that after a few minutes of petting a dog, the levels of oxytocin was higher in both the dog and her admirer.

Couple relationships are also affected by oxytocin. In a study conducted by scientists at the University of Zurich, forty-seven couples between the ages of twenty and fifty were videotaped as they engaged in one of their typical arguments. Just before the experiment began, couples were given either an oxytocin nasal spray or a placebo spray. The level of cortisol was measured in their saliva. Couples who were high in oxytocin were much more likely to listen to one another attentively and to laugh affectionately, while couples who received the placebo constantly interrupted and criticized each other. As we can now guess, cortisol levels of the couples receiving oxytocin was lower. In a related finding, it's been shown that when someone in a stable romantic relationship becomes dissatisfied with his or her partner, the levels of oxytocin increase, possibly to promote increased bonding and correct the situation.

While men also produce oxytocin, the analogous male hormone is vasopressin, involved in erection and ejaculation and also associated with social behavior such as aggression. Vasopressin reduces a man's inclination to perceive a face as friendly and promotes the likelihood he will see anger and threat even in an unthreatening situation. When oxytocin is given to men, they are able to more accurately identify emotions on faces. These oxytocin-influenced reactions and an emphasis on feelings may go a long way toward explaining why, as the best-selling book declared, it seems that "Men Are from Mars, Women Are from Venus."

The hormones oxytocin and vasopressin have existed for more than seven hundred million years and are found in worms and insects as well as in people. They support our capacity for bonding and reproduction by calming down our fear centers, subduing our judgment (making us amorous even if our frontal lobes tell us we're choosing the wrong partner), thus brewing an instant cup of trust.

The paradox of being human is we have the best chance of our children surviving if we mate with someone who is genetically unrelated and whom we would thus instinctively avoid, for they would not be members of the "tribe." The fear that needs to be overcome, so crucial to the survival of the species, is snuffed out by the firehose spray of oxytocin that accompanies young love and, all too often, leads to misplaced trust and suspended judgment.

Stress in Men and Women

Tiptoeing around brain differences between men and women does both a disservice, but acknowledging them can create quite the furor, as I learned to my chagrin.

A few years ago, I was talking to a colleague on the phone about a new patient referral.

"I'd like to meet you one of these days," I said.

"We've already met," she replied. "You actually made me very upset. I don't know if you remember, but you gave a talk where you said that men were better at math and science than women. I thought, what a horrible thing to say! We need all the help we can get, and here you were shooting us down."

Boy, do I remember this talk! I'm still getting flak for it nearly three years after delivering it at a luncheon benefit for the Lenox Hill Hospital, where I am an attending physician. The gist of the talk, in fact 99 percent of it, was about what occurs in women's brains with menopause. In the course of describing these changes, I happened to mention that, on average, the brains of women were superior to the brains of men in language tasks and that the brains of men were superior to women's when it came to mathematics.

At this luncheon at New York's famed Rainbow Room—where, if I remember correctly, nearly all of the men in attendance were part of the catering staff—I created a small tempest. Even though I mentioned Larry Summers, the disgraced president of Harvard, in the prelude to my observations, I had only a small inkling of what he must have faced in the firestorm sparked by his pronouncements about men's and women's brain differences.

I defended myself, noting the long line of talented professional women in my own family. My paternal grandmother was headmistress of a high school and was more educated than my grandfather. My mother was a microbiologist who left her young kids with babysitters and sometimes on their own as she pursued her (no doubt) stressful career.

Like oxytocin, the stress hormone cortisol is produced by both sexes. But the reason for its activation differs markedly in men and women. Women seem to be more adversely affected by interpersonal circumstances, such as social rejection or disconnection, and their cortisol production is related to that fact. Moreover, their predisposition to anxiety and mood disorders may be more apparent in an increasingly urban and less personal world, as women suffer from the lack of social ties more than men do. Men, on the other hand, respond more to challenges to their competence or sense of success.

Psychologist Laura Stroud and her colleagues studied twenty-seven men and thirty-one women between the ages of seventeen and twenty-three, measuring cortisol response to stress. Participants were put through both achievement and social stress tasks. For the former, they were asked to solve, under time pressure, twenty-four extremely difficult addition problems. Then they were asked to memorize a twenty-line passage from Milton's *Paradise Lost.* For the social stress task, trained undergraduate confederates excluded or rejected the participant during two interaction segments, using a variety of gradually escalating verbal and nonverbal techniques, so that the procedure appeared unplanned. While there were no sex differences in mood ratings, men responded with greater cortisol activation to the achievement challenges, while women showed greater cortisol responses to social rejection.

There are also differences between men and women in which parts of the brain are activated by stress. In an elegant study at the University of Pennsylvania, the neuroscientific effect of stress on gender was evaluated in sixteen men and sixteen women in their twenties. All subjects had their brains imaged at baseline, again during two levels of stress testing, and then after the tests were completed. During the high-stress test, they were asked to perform difficult arithmetic tasks. During the low-stress task, they counted back aloud from one thousand without any pressure. In addition to reporting to the investigators how stressed or anxious they felt, the subjects underwent brain scans and gave samples of their saliva so that their cortisol levels could be measured.

All the subjects reported an increase in stress and anxiety, and both men and women had increases in heart rate and cortisol production during the tasks. However, there was a definite gender difference in the areas of the brain that were activated by the stress tasks, offering clues to women's predisposition to anxiety. In men, stress activated the parts of the frontal lobe that deal with judgment, and it reduced activity in the parts that are more emotionally involved (left orbitofrontal cortex and inferior frontal cortex). In women, the parts that were activated were the insula and the areas of the reward system, as well as the middle parts of the frontal lobe—all regions more involved in emotion. Even as different areas of the brain were activated in response to stress in women and in men—and women reported more stress and anxiety—men and women were equally successful at solving the problems at hand.

The activation of more emotional brain circuitry in response to stress in women would have been a good evolutionary coping strategy. The tend-and-befriend pattern reflects the fact that women have to nurture offspring and bond in social groups in order to survive during crises, whereas men have to confront the stressor and make a decision to either overcome the predator or flee. So although both men and women may activate the sympathetic nervous system and secrete cortisol during acute stress, after the stressors are removed, men's and women's responses are different. Women activate those brain networks that increase attachment and caregiving and that subdue the activity of the sympathetic system and the production of cortisol. For men, responding to the predator demands that they increase alertness and focus, which are features of the prefrontal cortex.

By activating the brain's reward system in response to stress—where there are receptors for feel-good hormones like oxytocin and endorphins—women reduce the fight-or-flight response. And this helps them cope with long-term stress. Even so, by activating emotional brain circuits, women are at higher risk for emotional reactions, including anxiety. Also, women are slower to reduce the cortisol in their bodies after a stressful event, and the persistence of high cortisol levels may put women at risk for anxiety and depression.

To complicate matters further, women's customary cognitive style makes them more at risk for anxiety. They are more likely to ruminate, to focus on symptoms of distress, thus activating deeper, more midline parts of the frontal lobes, areas important in processing emotion. This is what usually makes women better listeners than men (as well as empathic doctors). But repetitive thinking and deep frontal lobe activation may predispose women to have more melancholic thoughts, more anxiety, and more suicides.

Even though the stress response in women activates the emotional circuits more than in men, this does not mean that women are more "emotional" and men more "rational" in the face of stress. Nor does it suggest that there are consistent or significant differences in intellectual ability. Keep in mind that in the University of Pennsylvania study, the men and women were equally able to complete the tasks. That said, while most of us believe that men and women are capable of meeting the same challenges, there are tasks—both mental and physical—that women may be naturally better at.

The authors of the University of Pennsylvania study point to possible gender differences in math ability by stating, "Females may feel more threatened by the arithmetic task than males." Whether or not women feel threatened is not clear, but when it comes to spatial and mathematical skills, we do know there are documented differences between the genders. For example, men steer using geometrical coordinates to gauge distance while women use landmarks to find their way. This difference is seen in rodents, too: Male rats use directional and positional information to negotiate mazes, while female rats use landmarks. It has not yet been shown that male rats are less likely to ask for directions. Perhaps it is this geometry-driven bias for directions that predisposes men to rely on their Global Positioning Systems to "recalculate" even as they drive by the clearly marked signposts.

These stress studies of men and women also have implications for gender differences in learning. When female rats are exposed to stress, they have more damage to brain memory circuits than male rats do; and females grow more connections in this circuitry than males do when they are placed in enriched environments conducive to learning. It appears that males learn better under stress, while females learn more in nurturing environments.

Older, Wiser, Networked Women

I stand by my statement at the Rainbow Room about math and language skills and the differences in brain strengths between men and women. In fact, it is in some measure a woman's exceptional verbal ability, among other things, that allows her to reach her career heights in her forties and fifties. However, as I told my luncheon audience, when a woman enters menopause and her estrogen levels plummet to zero, her brain seems to go haywire. She becomes more anxious and less calm, partly from losing sleep thrashing away at night with hot flashes, and also from hormone-influenced brain changes that may affect her ability to speak, remember, and react to stress.

According to research I conducted in 2001, women complain as much about memory loss—which invariably makes them anxious—as they do about hot flashes. Over and over again in my practice, I hear women lament that their "minds have left" them. Here's one such woman, Allison, talking about her menopausal "mind melt."

"I used to be the smartest, most verbal person on our team," said Allison, a fifty-four-year-old human resources manager who consulted me because she was afraid she had Alzheimer's disease. "When someone didn't remember something, they came to me, Allison. I was like the human computer, the steel trap. Nothing ever escaped me. I could type and text and answer the phone and problem-solve and manage my office staff without breaking a sweat, all at the same time. Now I'm a babbling idiot sometimes. I feel as if my brain is melting down.

"If it wasn't for the search function on my phone, I swear I couldn't handle my life. I may remember someone's job and look them up that way. In my contact list, I categorize people several different ways—for example, 'Dominick, physical therapist, arthritis care, Upper East Side'—just so that I can look them up any different way because I'm not sure what I can remember.

"Words leave me, and I've always had a fantastic vocabulary. Sometimes I feel I'm more demented than someone with Alzheimer's disease. Sometimes I think I actually have Alzheimer's."

When the *Wall Street Journal* published an article on menopausal memory problems, my office was deluged with phone calls from women like Allison, who feared they were going out of their minds. It seems a cruel trick of fate that at the time when many women are at the peak of their careers, the very tools that have helped them achieve their success begin to fail them. Estrogen or lack of it also affects mood, as anyone who has experienced premenstrual syndrome (PMS) will attest. In my practice, the judicious use of hormone replacement therapy has let me ease these symptoms in many patients. And for women who do not want to take hormones, I find that a time-limited course of antidepressants helps their brains transition into functioning without estrogen.

Menopause and the peculiar trials this transition poses are a problem specific to modern times, since even a hundred years ago, most women were dead by age fifty, before menopause could compromise their composure. In addition to its physical and mental stress, menopause is a looming shadow for many modern women, who, more than any other generation before them, are pushing parenthood out to the very edge of possibility. The decision whether or not to have a child can create tremendous anxiety, especially as women approach their forties. The prospect of not being able to conceive and perhaps having to face the stressful ordeal of fertility treatments is another impediment to calm.

For many women, the tick-tock of the biological clock is as terrifying as Poe's beating heart, sounding the death knell of their mothering dreams. I suggest to women wrestling with this decision that they freeze their eggs, so that in the future, at the right time, they can still opt for biological children. I find that in my practice, young women feel relieved and noticeably calmer once we talk about this solution.

Menopause, of course, is not the only stressor for women professionals and executives in their forties and fifties. By this time, many women have reached a high point in their careers, and a significant number of them are managers. And with leadership often comes increased stress. My colleague the cardiologist Marianne Legato has pointed out that women bosses generally get upset if someone in their office is angry with them. I myself feel discombobulated when all is not hunky-dory with my staff. Men, on the other hand, are rarely as affected by workplace discord. As we've seen, women need support systems more than men do.

Regardless of how much care and encouragement we receive from others, being a leader is a lonely job for both men and women. And women leaders face particular challenges, both from within (your brain) and from without (your staff). All leaders are often alone at the top, but a woman's brain may respond with more anxiety to this loneliness than a man's. The brains of women, their hormones, and their oxytocin levels, all speak to the importance of maintaining supportive social networks for calm.

As for coping with staff challenges, the female manager may be more likely to build consensus and pay attention to staff input than her male counterpart. Despite this cooperative leadership style, female managers may be judged more harshly and be less tolerated for their foibles, by both their male and female staffers. And this can make for more anxiety in a female leader.

Given these harsh judgments, it shouldn't surprise us that many people—men and women—would rather not work for a woman, and report more work dissatisfaction when they do. Ninety percent of female Dartmouth MBA students said they preferred a male boss. A study reported in *Forbes* found that women working either for another woman or for a male-female team had higher stress levels than women who worked for men. The article sparked a robust response from readers. One woman wrote, "We women are very competitive with each other . . . emotions and feelings get in the way." Indeed, women under stress are more likely to recruit their "emotions and feelings," even though their performance does not suffer.

In a British study of sixty men and sixty women managers, women reported more pressure from their jobs than men, and women in male-dominated industries reported worse mental health and more stress when they used interpersonally oriented leadership styles, while men had no such difficulty.

Two Roads to Calm

The stress surrounding gender does not end at the office. Meet Gabrielle, a forty-four-year-old MIT graduate, wife, and mother of two, who quit her job in finance to spend more time with her kids.

I met Gabrielle at a dinner party given by a mutual friend. Here's how she described her situation to me: "I want to be respected, to be able to offer something. My husband thinks I'm just a housewife; he treats me like I'm a nanny. I want to tell him that I'm doing something important, but he doesn't respect what I'm doing. When I was working, I brought home a big salary; I had stories to tell about the big finance guys I worked with. Now I just feel like, what do I have to offer? Some of my friends are amazed that I'm doing the twenty-four-seven mommy thing. I love my kids, and I know taking care of them is important work, but I worry I'm not interesting anymore. I'm even boring to myself. I'm afraid I'm boring you as I talk about this. I used to be the life of the party; now I just don't feel I have much to contribute."

Gabrielle has opted out of the workplace rat race, at least for now. Sadly, though, giving up her career has led her to lose some of her husband's respect and even her own self-regard. In our era of embattled "mommy wars" and office infighting, women face tough choices and can find emotional equanimity hard to attain.

Women today, whether they're managers, full-time mothers, or professionals, whether they're sixteen or sixty, face unique challenges in their quest for calm. For one thing, right from the start, women are more prone to anxiety and depression and more likely to be medicated for these ailments. We need to be aware of this. We also must honor and appreciate the fact that women achieve calm through sisterhood and through mediation, as opposed to aggression. We have to acknowledge that women, more than men, need friends to help us reach a state of calm, and that we need to give and receive social approval.

It's clear that men and women recruit different brain regions and strategize differently when trying to get from here to there. When they're searching for the nearest 7-Eleven, women use landmarks, and men use navigational signs, but we must recognize that men and women arrive there at the same time. Embracing differences, rather than denying them, helps us understand the powerful distinctions—and similarities—between the sexes.

CHAPTER 9

Calm Parenting Is Possible!

Young Paul, plagued by his cold and beautiful mother's unspoken anxieties about money, desperately wants to help her. Frantically riding his rocking horse day after day, he finds his luck, acquiring the magical skills to pick winners at the horse races. His uncle places the wagers and passes the monies on, anonymously, to the mother. This is the setup of a story by D. H. Lawrence titled "The Rocking Horse Winner." The climax is gut wrenching:

> The room was dark. Yet in the space near the window, she heard and saw something plunging to and fro. She gazed in fear and amazement.
>
> Then suddenly she switched on the light, and saw her son, in his green pajamas, madly surging on the rocking-horse. . . .

"Paul!" she cried. "Whatever are you doing?"
"It's Malabar!" he screamed, in a powerful, strange voice. "It's Malabar!"

And so Paul correctly picks the winner of the derby, assuring the family's fortune. But he dies a few days later, exhausted by his efforts.

While the notion of a little child riding a rocking horse to death to make his mother happy is fantastic and overdramatized, we all have seen children do nearly anything to allay their parents' troubles, even something that seems preposterous to an adult.

I read Lawrence's story when I was quite young, perhaps six or seven. It struck such a strong chord in me that, as I prepared to write this chapter on parenting and calm, that story from many years ago—and the long-ago emotions it evoked—immediately came to mind. My childish core brain had resonated with Paul's pathos and empathized with his predicament, his worry about the fate of his family and his determination to change their luck. Paul's mother epitomizes the extreme of anxious parenting for me. So deeply unhappy and unsatisfied, so fixated on her bad luck, she could not be emotionally available to her children. To her sensitive son, she transmitted only her sorrow about being insufficiently rich, so much so that he gave up his life trying to please her. Lawrence himself had a difficult childhood; he was one of several children of a barely literate coal miner and his teacher wife, and it is easy to imagine him caught up in the storm of their marital discord and financial woes.

In the real world, anxious parents are more likely to raise anxious children. And modern children, thanks to rampant media exposure, are more anxious than ever, with eight- and nine-year-olds fretting about the state of the world. Even if parents responsibly limit their kids' exposure to the Internet and television, children can't help but overhear them talking worriedly about the latest terrorist attack or rebel uprising.

"That's nothing for you to be scared about," said one of my patients to her young daughter, referring to a random act of violence that left several schoolchildren dead. Of course, her daughter could not help being aware that the children killed in the attack were exactly her age. But the mother's reassurance may have been too vague for this nine-year-old. More helpful would have been specific reassurance, such as, "Yes, what happened at that school was terrible and scary for you. At your school, we as parents and your teachers make sure that students don't carry guns into class." This more pragmatic answer concretely addresses the child's concerns and is more soothing to the core brain.

Then, too, children pick up on unvoiced parental anxiety, much as Paul did in Lawrence's story. Children often react as strongly to what parents don't say as to what they do say. Body language is a potent communicator. Lawrence tells us: "Everybody else said of her [the mother] 'She is such a good mother. She adores her children.' Only she herself, and her children themselves, knew it was not so. They read it in each other's eyes."

A mother can transmit anxiety to her offspring even before the child is born. Much as I hate to malign mothers after a century of maternal blaming instigated by Freud, it is known that anxious pregnant mothers may be more likely to have anxious children, for the influence may be felt even in the womb. In a seminal study, British scientists tested cortisol levels in 267 pregnant women and the amniotic fluid surrounding their fetuses. They found strong correlation between these levels, with a higher stress-derived cortisol in the mother translating into higher cortisol levels in her fetus. This may explain why maternal stress not only causes lower birth weight but also alters brain development, which may be one reason why such babies are prone to grow into anxious adults.

Another study, in Belgium, measured anxiety levels in seventy-one normal mothers and their firstborn children when the children were between eight and nine years old. The researchers found that the mother's anxiety level during pregnancy, particularly between twelve and twenty-two weeks, was a predictor of anxiety in her children. Interestingly, anxiety later in the pregnancy—thirty-two to forty weeks—did not seem to engender anxiety in the children. Why was this so? The answer lies in the so-called "fetal programming hypothesis," whereby maternal anxiety or another disturbance at a crucial time during pregnancy is associated with changes in fetal brain development. The core-brain fear center and much of the brain develops by the time the fetus is twenty-four weeks old. If the mother is anxious during this critical neurodevelopmental period, her emotional state is transmitted to the fetus's developing fear region, among other areas, and may put the child at risk for anxiety later in life.

The good news is that even when children are predisposed to anxiety in the womb or as a result of their parents being anxious around them when they are growing up, such natural anxiety can be modified by the right kind of nurture. Parents need to talk to their kids about crises and even about everyday problems in concrete and simple ways, suited to the child's age. They can reassure the child realistically that things may be bad right now but that they have a plan to improve them or even just to deal with them. "Daddy's looking for a new job; we have money in the bank to help pay the bills, but we need to spend less. So no fancy presents this Christmas" is better than "Don't worry, everything will be okay." "Mommy is sick, but the doctors are helping her get well" is better than "Mommy's not sick at all. I'm just tired, running around after you, little guy!" An older child will need to know more—"Daddy is training for a new type of job." "Mommy has breast cancer, and she is getting the very best treatment."

Denying or ignoring a problem does not help children, because they have fantastic imaginations and will invent scenarios far worse than the reality. The younger the child, the more sensitive he or she is to nonverbal cues. The house doesn't have to whisper about money for even a toddler to pick up on his parents' nervousness about finances. His core brain senses the problem at a primitive emotional level. And then, because the information is not tempered intellectually by mature frontal lobes, it is amplified to rocking-horse proportions. Anxiety and a lack of calm are inevitable results.

Not long ago, both the extent of a certain family's denial and the futility of trying to cover up bad news were bought home to me in a dramatic way. A patient of mine, Florence, was only forty-eight when we diagnosed her with early Alzheimer's disease. Ever the devoted mom and able college professor she once was, Florence wanted the diagnosis kept from her three teenage children at all costs. I long have been a firm believer in not hiding important information from children—and certainly not from adolescents, who, despite their still immature frontal lobes, are certainly capable of understanding even terrible news like this. So I was determined to help the family find a way to deal with Florence's condition. After much cajoling from her husband and gentle persuasion from me, Florence agreed to a family meeting so that we could discuss the diagnosis and implications with the children.

I approached the meeting with caution. Florence told me that her son and daughters had no inkling of why they were at the meeting other than that they were here to see their mother's physician. Once we were all gathered in the room, I began gingerly.

"You must be wondering why we are meeting today," I began but was cut off immediately by the thirteen-year-old, Patrick.

"You want to tell us that Mom has Alzheimer's," he piped up. His older sisters tried to shush him, but he was having none of it. "We knew a long time ago that she wasn't right."

"Yeah," agreed the older sister, somewhat sheepishly. "We looked it up online and figured out that's what it was. We didn't want to upset Mom with what we knew 'cause we knew she'd worry that we knew."

You could have dropped a pin in the next room and we'd have heard it. As I say, it's hard to hide big news from children—and it only adds to their anxiety if you try.

How common is anxiety in children? It turns out that it's more prevalent than all other childhood mental illnesses clumped together. A staggering 17 percent of eleven-year-old boys and girls are classified as clinically anxious. More than half the adults being treated for anxiety report feeling anxious as children, suggesting that anxious children are likely to grow up to be anxious parents.

Even if parents are calm or manage to feign composure in front of their children, we must keep in mind that children bring their own innate levels of calm to every setting and situation. Children are born with particular temperaments—relaxed, edgy, aloof—that tend to persist through their lives.

A Child's Vagal Tone

The Harvard psychologist Jerome Kagan, famed for his innovative and influential research on child development, has followed groups of children from early infancy through adulthood. Looking at their behavior at age two, he has been able to predict which children will turn out to be anxious and which ones will grow up to be calm. Kagan not only was able to forecast these temperamental styles into adulthood, but, even more impressively, he was able to tie them to vagal tone, the basic tenet of our bottom-up theory of calm.

In the early 1980s, Kagan evaluated forty-three children aged twenty-one months and monitored them for two years, correlating shyness and behavioral inhibition with vagal tone. In a recent e-mail, he told me that he measured heart rate and variability of heart rate in response to stress, assuming that "a high-reactive temperament was due to an excitable amygdala, which sends projections to the sympathetic nervous system." He pointed out, too, that many other researchers "have found that anxious adults are a little more likely to have higher heart rates combined with lower variability."

During these evaluations, an interviewer examined each child for over an hour and the interviews were videotaped. In addition to the child's responses to various commands, such as sitting still for thirty seconds, and performing cognitive tasks, the child's response to stories provoking empathy were also observed. The children's responses to their mothers and to the examiner were monitored, along with their heart rate, using two stick-on leads. Changes in heart rate and heart rate variability from beat to beat with various tasks were also measured. Finally, when the interview was finished, each child's ability to engage with another unknown peer in play was observed. In this way, Kagan and his group thoroughly documented many aspects of anxiety, sociability, and its relationship to vagal tone in children.

As they turned four, Kagan's twenty-two inhibited kids continued to be more fearful, less interactive with their peers, and less adventurous than their more gregarious peers. And here's the kicker: The twenty-one confident kids had higher vagal tone to begin with, indicating that the children brought their own innate temperament to the equation.

Kagan has studied these children into early adulthood and has continued to confirm these findings. But this does not mean that children born with low vagal tone cannot be helped. Indeed, they can benefit from a calm parenting style. In research designed to address this issue, Kagan and his colleagues evaluated vagal tone in 104 two-year-olds and assessed their mothers' parenting styles. Two years later, he reassessed the entire group of mothers and children. Both vagal tone and parenting practices appeared stable over that time. Not surprisingly, the parents of those children who were irritable and fearful—that is, those with low vagal tone—were more likely to report restrictive parenting practices. In other words, anxious children evoked in their mothers a parenting style that made the children more anxious.

Although many psychologists have studied children's temperament, Kagan is unique in tying it to vagal tone. Moreover, the longevity of his research, which spanned several decades, and the level of documentation, using videotapes as well as direct observations, are unprecedented. His findings have provided a rich picture of mother-child interaction and shed new light on how anxiety develops. Irritable infants who made their mothers respond more punitively were even more angry and poorly behaved than irritable infants whose mothers responded supportively. Toddlers who are shy and anxious to begin with and who have overprotective and intrusive parents are even more likely to be withdrawn as they grow up. Calm parenting is particularly important for such children.

Is the reverse true? Does low vagal tone in parents predict more anxious children? Psychologist Susan Perlman and her colleagues studied forty-two children aged four to five, most accompanied by their mothers, three with their fathers, one with a grandmother, because these were the primary caregivers. Vagal tone and emotional responsiveness were evaluated in the parents, and anxiety in upsetting situations was measured in children by showing them cartoons that depicted stressful situations. Parents with high vagal tone reported better home environments and calmer emotional responses to stressful situations. The children of parents with high vagal tones were more adept at dealing with emotional situations. As in all of nature, the interaction works in both directions.

Finally, is vagal tone inherited? We know that anxious parents beget anxious children through both genetics and nurture, but does low vagal tone beget low vagal tone? The answer is not yet known, but we do know that certain aspects of heart rate physiology are inherited. So my guess is that some aspects of vagal tone are in fact heritable.

Calm parenting of an irritable child with low vagal tone can be a bit counterintuitive. He or she will more likely grow up calmer if you address your parenting toward increasing the vagal tone rather than trying to eliminate the annoying behavior. The best way to help these children manage their emotions and behavior is from the bottom up—holding them more, being more physically soothing, and imposing fewer restrictions. Some rules are of course necessary, but the point is that achieving better behavior through better vagal tone is more effective than trying to reason with these irritable or fearful children. The top-down approach of talking and arguing falls on the deaf ears of their still immature frontal lobes. Physical comforting, from holding to stroking to using soothing tones to whispering sweet nothings in a calm voice, benefits all children, not just those with low vagal tone.

Even as parents try to comfort and support their children, sometimes that support is misplaced or not adapted to the child's age—and it causes more anxiety instead. In a café recently, I observed a mother with her young son, who looked about five years old. They walked in for breakfast on a crowded Sunday morning and found a small corner table set for four people.

"Where do you want to sit, honey?" asked the mother, pointing to the four chairs.

"I don't know, Mom. Wherever," replied the boy, his voice still heavy with sleep.

"You can sit up against the wall; you can sit in the corner. Or you can sit next to Mommy, right here," said his mother, ignoring his indifference. "If you sit next to the wall, you can watch people come and go. What do you want to do?"

"Mom, I don't care," said the son, starting to whine now.

"Okay, but don't cry about it after," warned the mother. "Didn't you want to draw with your crayons?"

"Uh, okay. Here?" The son gestured toward the corner seat.

"Good," said his mother, looking pleased that he had made a decision. "Now, what would you like for breakfast?"

And so it went. By the time they left the café, this five-year-old had had to make multiple decisions, choosing from a variety of options, with regard to his seating, his breakfast, his choice of toast, his choice of coloring book, his other toys. I was exhausted. I sometimes thought my mother was too strict, but watching this sad Sunday morning drama, I was glad she set down a plate in front of me for breakfast every morning and that I ate what was on it. We had little choice of seating. A good thing, I now realize. Making the "right" (or even the "wrong") choice at a young age would have been stressful for me, or for any child.

As I mentioned earlier, Barry Schwartz has argued compellingly that choosing from an array of possibilities is no picnic even for adults. The overwhelming number of choices available today—from breakfast cereals to TV channels—places a burden on our brains. Now imagine the same choices when your frontal lobes, the final arbiters of choice, are not even close to mature. In fact, they don't develop fully until your mid-twenties.

Children use core brain and the bit-player regions of the brain in making decisions much more than they use their rational frontal lobes. Skill sets such as driving or playing the violin may be learned early, but choosing not to drag race a high school buddy over a barren stretch of highway is not always tempered by the rationality of the frontal lobes, as highway mortality statistics attest. This is not to say that activities like chess and other logic-driven games should not be introduced and fostered at a young age. Such games recruit frontal lobe resources and are best cultivated at a young age, but the type of frontal-lobe involvement is different. Imaging studies reveal activation in the frontal lobes during the strategic planning that is crucial to playing to checkmate. Children who pick up chess at five or six play the game almost intuitively, aware of the rules without any conscious thinking, for their brains have been "molded" by the rules of chess.

However, the difference between choosing what move to make in chess and choosing where to sit at the breakfast table is vast, particularly in the mind of a child. Chess has rules that can never be affected by emotion and all the nuance of human language, whereas everyday choice, even involving something as mundane as breakfast-table seating, with its myriad permutations and computations, has no such rules, and winners and losers have no absolute definition. Uncertainty is the mother of anxiety.

What Parents Can Do

When it comes to discipline, parents also need to take brain development—and vagal tone—into account. Grounding your out-of-control daughter by sending her to her room, where she can continue on Facebook, AIM, and texting her friends, does not calm her brain. Putting your tantrum-throwing son into "time out" in a room full of computer games is no time out at all. It does not calm the vagus or the core brain—and you can count on another tantrum before long.

The prototypical time-out picture that springs to my mind is of a child in a rocking chair, facing a corner of a blank wall, as the parents sit quietly in another corner of the room reading. The child is safe with her parents close by, she is in a rocking chair, the motion of which is particularly soothing to the vagus, and she has no stimulus to distract her aside from the wall. That is a core-brain calming time-out.

In the increasingly overstimulated world that is the milieu of the modern urban child, there are too many toys, too much technology, and too many choices. The one thing that children don't have enough of, in my opinion, is community. And this is what the core brain requires. It needs the skills gained from living in communities to help it empathize and communicate effectively, which is particularly important for calm.

Overscheduling children's lives, in the race to Harvard that begins at conception, leaves little time for the insouciance of spring, one of the joys of childhood. Gone, sadly and probably forever, are the days of stoop ball and neighborhood pickup games on blocked-off streets. Impromptu playtime and other core-brain delights give rise not just to calm but to a productive and healthy adulthood. Researchers have in fact found that the more activities that are scheduled for children, the more likely they are to suffer from stress and anxiety. In fact, the rates of diagnosis of attention deficit disorder, conduct disorder, and depression are skyrocketing in urban children. And as we know, anxious children become anxious adults.

Accompanying frenetic scheduling is the constant need for vigilance. A patient of mine told me about her ten-year-old granddaughter who lives in a New York City apartment building. Although the building has a doorman who monitors the front door and all visitors, the child is not allowed to visit her friends on the third floor from her apartment on the seventh unless accompanied by a parent. What danger may lurk in the elevator? It is not uncommon for parents to monitor their child walking down the hallway to visit another child's apartment. What danger might lurk in the hallway? This kind of anxiety about unseen dangers surely has an impact on the tenderly impressionable core brains of young children, possibly making them more suspicious or anxious in general, and surely less calm.

Our children's core brains and their brain fear center, with its inherited, evolutionarily based fear of strangers, produces an indiscriminate sense of anxiety, of vigilance. This can have a lasting effect on the rapidly developing frontal lobes, which may then be prone to more suspiciousness of strangers throughout life.

Related to overvigilance and lack of community is the proliferation of technology, which further handicaps children in their quest for calm. Computer games (especially the handheld ones found in every child's backpack these days), with their beeping and flashing and demand for tight electronic focus, further compromises community and core-brain calm. As does children's texting. Every parent has had the infuriating pleasure—or soon will—of trying to compete for their children's attention while the kids tap away on their tiny keyboards. Texting impairs the honing of core-brain people skills. As a boy who flirts through texting with a girl he adores said to me in despair, "I don't know if she likes me in real life!"

Unfortunately, parents are just as guilty of this. How often do we see a child in a park or at a restaurant trying to catch the attention of a parent who is talking on the phone or texting with abandon? When this goes on constantly—as it often does—children may become less adept socially, making it harder for them to communicate with others.

It is difficult to be calm when you haven't developed the skills to deal with your fellow human beings. Extreme examples are children with autism, many of whom suffer from severe anxiety and an inability to self-soothe. Psychologist Stephen Porges has worked extensively in the area of vagal tone and core-brain sensitivity to both emotional aspects of communication and the impairment of this circuitry in patients with autism. He contends that the vagus is intimately engaged with core-brain circuits that are involved in the expression and recognition of facial expressions, so that impaired vagal tone will also lead to impairment in nonverbal, emotional communication with others, as is the case in autism.

Just as social anxiety impedes calm, a childhood steeped in excessive vigilance, even at the hands of well-intentioned but overly protective parents, may impair one's ability to self-soothe, to self-protect. I was walking down Madison Avenue on New York's Upper East Side a few weeks ago, when I heard a mother say to her two-year-old child slouching in his stroller, "Aaron, close your eyes, it's getting sunny!" And little Aaron obediently shut his eyes, to protect himself from a gloriously sunny day because his mother thought the sunshine would harm him.

This example is funny in its absurdity, but it's no joke

that many kids today are the unwitting victims of their parents' germ phobia and oversanitization—of children and of childhood. The truth is that exposure to a reasonable number of environmental pathogens at a young age is helpful in developing immunity to a host of illnesses in adulthood. Preventing this exposure can set the stage for later susceptibility to illness. Let's face it, being a kid is all about the wonder of stepping in puddles and poking around in the ground and playing with other kids in real time. Overprotection can bring on real danger: those unnamed, unvoiced fears that resonate through the house, whispering into a child's ear, "Be afraid!"

I make no claims to being an expert when it comes to children, and only my daughter can say how calm a mother I am. But it seems to me that effective parenting must take into account the neuroscience of calm and the neuroscience of a child's brain, in order to be effective. Understanding the effects of the vagus on the core brain, the evolution of the frontal lobes and their role in anxiety, and the ability of a child to respond to unspoken messages will help guide parents in raising calm children, regardless of whether their children come into the world with low or high vagal tones.

Chess and piano lessons are all very well. But a calm brain is the greatest gift a parent can cultivate.

CHAPTER 10

Weaving a Tapestry of Calm

"Without a theory, the facts are silent."

—Friedrich Hayek, Nobel Prizewinner in Economics

Imagine yourself lying on a couch, in a warm, quiet, and soothingly lit room, talking about anything at all without fear of disapproval, secure in the knowledge that at the end of this process, you will feel better. Imagine that for forty-five minutes in the middle, beginning, or at the end of a hectic worry-filled day, you can count on this judgment-free unburdening of all that was a source of anxiety in your day. Imagine, too, that someone is listening to you with attentive devotion, someone endowed with curative powers, an omniscient being.

Welcome to "free association," one of the key tenets of the much-maligned practice of psychoanalysis. In the late

1800s, Sigmund Freud and his colleague Josef Breuer developed this technique to treat people who sought help for anxiety or other psychological woes. The premise of psychoanalysis is that in this safe space, without the constraints of the rational frontal lobes, patients will freely tap into their unconscious and release and process trapped memories, emotions, and thoughts, eventually curing their underlying illness.

Even as psychoanalysis and free association have since proven to be poor and cumbersome treatments for most types of anxiety, some aspects of the method turn out to be excellent promoters of calm.

Let us for a moment put aside Freud's theory of accessing the unconscious and focus instead on what we know about creating calm from the bottom up. Lying on a couch in a quiet room? Good for the vagus nerve and the parasympathetic relaxing system, as well as for subduing your sympathetic system's flight-or-fight reaction. Being in the presence of an authority figure whom you believe in and trust? A great way to turn down the judgment centers of your frontal lobe. Feeling safe, warm, and comfortable? Your brain's fear center is tuned down. Feel listened to and believe that you are in true communion with another human being? Your brain's "gut feeling" and "reward" centers are amped up. No wonder Freud's therapeutic brainchild is a practice that can be addictive. Even now, there are many loyal devotees who swear by psychoanalysis and especially free association. They say it's like meditation under guidance.

And indeed it does share similarities with meditation. It addresses core-brain needs, increases vagal tone, turns down amygdala and frontal cortical judgment activity, and promotes a sense of well-being. In throwing out the psychoanalytic bathwater, we must take care not to toss techniques that calm the core-brain baby.

The Upside-Down Brain

Soothing the core brain, whether with Freud's couch or with an inversion table, is an important part of the bottom-up theory of calm. I have reviewed many of the facts that attest to the importance of the vagus and the core brain in achieving calm, but as Friedrich Hayek the economist put it, "Without a theory, the facts are silent." Where do we go from here?

Applying the bottom-up theory of calm was effective for my patient Lisa, she of the bridal floral business whom you met in Chapter 1. Lisa has continued to do well; her crippling migraines, irritable bowel syndrome, anxiety, and neck pain have not come back. When I last saw her, she told me, "Now when I get anxious, I am not anxious about the anxiety. Before, I would get anxious and say I'll take something. I don't do that anymore. I use the table maybe once or twice a week, but I find I don't need it as much anymore. In truth, my attitude is fine. Every once in a while, I'm in a black mood. But I don't have as many downs. When I get a call from a client, and I start helping her plan her flower arrangements, I'm as happy as a clam!"

Unlike Lisa, some people may not realize what they are doing to calm themselves down. A Connecticut hedge fund manager, Jack, came to see me for tingling in his hands, which turned out to be carpal tunnel syndrome. I remarked on how fit he was, how despite the pressure of overseeing nearly two billion dollars in assets, he appeared relaxed and in good spirits. What was his secret? I inquired.

He said, "I think that it's the exercise I do every morning; it grounds me. I usually get to bed early; my wife likes the kids in bed by eight and we tend to have a quiet social life. I wake up around five or so, haven't set the alarm in years, just wake up around then. While everyone is still asleep, I go through my routine. Free weights, some yoga, some stretching, some treadmill work."

So far, so good, I thought. This sounded like a perfect recipe for calm: good sleep, good biorhythm, good family ties, good exercise. But then Jack dropped the bombshell. "I finish it off with about ten minutes on the inversion table."

I was stunned. "What? An inversion table?"

"Yes," Jack said. "It's one of those things they sell for back pain; you hang upside down from it. I got it because my back had been bothering me and now I swear by it.

I don't have back pain anymore; but I still use the table because I feel it decompresses the spine and prevents pain starting. I do my crunches on the table. I find that I'm addicted to it. It makes me feel good. When I get off it, I feel like I've flushed my brain."

In many ways Jack, the hedge fund manager from Greenwich, is like my grandfather from India. Despite the high stress of Jack's job, he was able to maintain a high level of equanimity by following a quiet and soothing morning routine, just as my grandfather did. In addition, he harnessed the power of bottom-up calm by building the inversion table into his life.

It is possible to achieve a similar effect from the classic yoga shoulder-stand pose. Recall my frenetic patient Annie, who was amazed that she could empty her mind and "go blank" when she was in this position. Her vagus imbued her with calm from the bottom up. One difference between the inversion table and a shoulder stand is that the latter can compress the spine in the region of the neck, whereas as you hang from your ankles on the table, there is no pressure along the spine. The other clear advantage of the table, of course, is that you don't have to train to be able to use it as you would for the upside-down yoga pose.

The inversion table satisfies core-brain needs and turns down the activity of the frontal lobes. When you lie head-down on a slanted table, your body is automatically in a vulnerable position, and your core brain knows you would not voluntarily put yourself in this position unless you felt safe. You don't want to be upside down if you need to run from an attacker. Nor would you be likely to recline on a couch in a darkened room if you thought danger was lurking nearby, even in turn-of-the-century Vienna.

A bottom-up position changes blood-flow dynamics, slowing down your heartbeat and breathing and inducing the core brain to relax. The inversion table is primarily designed to ease back pain, as it did with Lisa and with Jack. But by recruiting the parasympathetic system and turning down the sympathetic system, it quickly creates a brain milieu that is biased toward relaxation and away from stress.

The inversion table, along with his calming exercise routines, nurtured Jack's core brain and helped him keep his cool when job pressures heated up. But what about more typical executives, whose lifestyles puts calm far out of reach?

Working Hard

Here is a day in the life of a CEO of an international financial firm—let's call him Clifford—whose psychiatrist allowed me to interview him for this book. Clifford was being treated for chronic anxiety after he suffered a heart attack while working out when he was a mere forty-eight years old. His cardiologist felt his type A personality put him at risk for more heart-related problems and referred Clifford to my psychiatrist friend. Clifford went to the psychiatrist reluctantly. As he said to me, "Who has the time?" But he has improved on medication.

I wanted to take a look at an average day in Clifford's life to see what the activity level of the components of his brain calm regions was at different times. At my request, he kept a diary of a typical day, which as it turned out was an especially stressful twenty-four hours.

Clifford lives in New York City with his wife and four children. He typically puts in extraordinarily long hours. While this may be a particularly over-the-top example, it is not uncommon for company leaders to work days that start at six and end around midnight.

Clifford's judgmental, rational frontal lobes and his fear center are on all the time, whether he's in the office or out at the soccer game with his family. In addition, whether he is in the park or at dinner, he's also on his BlackBerry putting out real and imagined fires and keeping the virtual lions at bay. These virtual lions are much harder for the core brain to deal with than real lions—both because the brain is wired to deal with real lions and because there are far fewer real lions than virtual ones. Our alarm system, which is programmed to go off at the slightest hint of danger, keeps us safe from the virtual lions. In the world of business, this same sensitive alarm system is constantly beeping and is hard to shut off.

Here's a day in Clifford's life.

Time	Description	Theme, Neural Region, Activity
5:30 A.M.	I jump out of bed when the alarm goes off, check my BlackBerry for e-mails, run on treadmill, take a few calls, watch CNN, take a shower. Kids just waking up as I leave at 7:00 A.M.	Deadlines! Vagal tone—low Fear center—high Rational frontal lobes—high
7:30 A.M.	Breakfast meeting with a big client. He's thinking of shifting his assets to another firm. Not sure if I talked him out of it.	Fear, Anxiety Vagal tone—low Fear center—high Rational frontal lobes—high
8:30 A.M.	In office, more than thirty e-mails from our six Asian offices overnight. I'd flagged the most important ones while I was on treadmill this morning. In the office, my assistant is with me taking notes. A few calls, double espresso twice, no time for breakfast.	Multitasking Vagal tone—low Fear center—high Rational frontal lobes—high
10:00 A.M.	Meeting with the PR firm head. A big negative story about to break about firm; may affect our stock price. A whistle-blower has apparently spoken about the practices of one of our senior executives, whom we recently let go at full pay. This is going to be bad.	Fear, Anxiety, Betrayal Vagal tone—low Fear center—high Rational frontal lobes—high
12:00 noon	Lunch meeting with Larry, who runs a consulting firm; the Four Seasons has the best food, but I couldn't enjoy it. Larry's brilliant and fun, but I was preoccupied with press problem. Could I take him into my confidence? Decided no. Got to handle this alone.	Fear, Anxiety, Loneliness Vagal tone—low Fear center—high Rational frontal lobes—high
1:15 P.M.	Canceled scheduled meetings and discreetly arranged emergency meeting with the top three division heads. Worried about whistle-blower. We discussed damage control. No solutions yet.	Strategizing Vagal tone—low Fear center—high Rational frontal lobes—high

Time	Description	Theme, Neural Region, Activity
2–2:15 P.M.	Met with a few kids from my daughter's school to discuss my work for school newspaper. Tried to cancel, but they were here during school hours; finished very quickly with them.	Community, Annoyance Vagal tone—low Fear center—high Rational frontal lobes—high
2:15 P.M.	Meetings. Drafted a letter to board about problem, spoke again with PR. We decided to lie low and wait.	War, Vigilance Vagal tone—low Fear center—high Rational frontal lobes—high
6:00 P.M.	Had promised my younger daughter I would be there for her soccer game; showed up in second half, stayed till end. She scored two goals, she's very athletic, but I missed goals as I had a bunch of e-mails on BlackBerry and decided to meet with PR guy one more time to finalize for tomorrow. Canceled dinner plans with family—wife and kids upset. Canceled doubles tennis with wife and friends of ours.	Community, Vigilance Mirror neuron's not attuned to children's needs [?], preoccupied. Vagal tone—low, Fear center—high Rational frontal lobes—high
8:30 P.M.	Finally home, kids working on homework, rough-housed with them a bit. Then they went off to bed. Still preoccupied with work.	Community, Vigilance Vagal tone—low Fear center—high Rational frontal lobes—high
9:45 P.M.	Spent some time with my wife; she was very upset, thinks I work too much, which I do. She went to bed angry. I stayed up to finish e-mails.	Loneliness, Anger, Vigilance Vagal tone—low, Fear center—high Rational frontal lobes—high
1:00 A.M.	Couldn't sleep so caught up on trade reading and also the *Economist,* which I find is excellent. Finally turned in, took an Ambien to help me unwind.	Drug-induced meager sleep Vagal tone—low Fear center—high Rational frontal lobes—high

One thing that stands out in this schedule is the inability of Clifford's frontal lobes and his amygdala fear centers to disengage at any time, even when he is with his family. That is because this area of the brain—the so-called simulator—which creates simulations of possibilities and deals with choices, is particularly robust in people in leadership positions. Take these folks and plop them on a beach chair in the Caribbean and they are like fish flapping on the sand. With no decisions to make, beyond what to eat for lunch, they take to their BlackBerries or ruminate incessantly, unable to calm down. Virtual lions lurk behind their beach chairs, and squalls threaten their piña coladas' paper umbrellas, keeping their brains on high alert.

As we saw with Winston Churchill, however, good leaders are able to relax, to truly relax. And building relaxation into their daily lives allows them to perform better.

Let's see what we can do with Clifford's life to make it healthier and more fulfilling to him and his family while still allowing for his professional edge.

Time	Description	Theme, Neural Region, Activity
6:00 A.M.	I generally automatically wake up around six, nuzzle with my wife for a bit. We talk about the day ahead, have a quick shower, eat breakfast with the kids. It's complete chaos with all four of them. Get picked up for work at 8:00 A.M.	Community, Oxytocin Vagal tone—high Fear center—low Rational frontal lobes—low

Time	Description	Theme, Neural Region, Activity
8:00 A.M.	I've very few breakfast meetings, intrudes on family time. Client was rescheduled for a lunch meeting.	Community Vagal tone—high Fear center—low Rational frontal lobes—low
8:30 A.M.	In office, more than thirty e-mails from our six Asian offices have been weeded and flagged by my able assistant, now that I have learned to delegate. We spend time on them, take a few calls.	Multitasking Vagal tone—low Fear center—high Rational frontal lobes—high
10:00 A.M.	Meeting with the PR firm head. A big negative story about to break about firm, may affect our stock price. A whistle-blower has apparently spoken about the practices of one of our senior executives, whom we recently let go at full pay. This is going to be bad.	Fear, Anxiety, Betrayal Vagal tone—low Fear center—high Rational frontal lobes—high
11:00 A.M.	End meeting early, need to think more about this. Work out for about thirty minutes, then take a quick shower, head for lunch.	Self-soothing Vagal tone—high Fear center—low Rational frontal lobes—low
12:00 noon	Lunch meeting at Four Seasons with Larry. He runs a consulting firm. Savored the food. Larry's brilliant and fun, enjoyed him. Could I take him into my confidence? Decided no. Got to handle this alone.	Community, Loneliness Vagal tone—low-high Fear center—low-high Rational frontal lobes—high
1:15 P.M.	Canceled scheduled meeting and set up emergency meeting of the top three division heads. Worried about whistle-blower. We discussed damage control, as well as the solution that dawned on me after lunch.	Strategizing Vagal tone—low Fear center—high Rational frontal lobes—high

Time	Description	Theme, Neural Region, Activity
2–2:45 P.M.	Met with a few kids from my daughter's school newspaper to discuss my work. Great kids, had a blast, took them down to the company cafeteria for ice cream and they interviewed me there.	Community Vagal tone—high Fear center—low Rational frontal lobes—low
2:45 P.M.	Meetings. Drafted a letter to board about the whistle-blower problem. Spoke again with PR; we agree on my solution of a one-page press release succinctly describing the facts. Decided that our quick response would be best for damage control.	War, Vigilance Vagal tone—low, Fear center—high Rational frontal lobes—high
5:00 P.M.	Left work early, because promised my younger daughter I would get to her soccer game; my assistant knows to field calls for the hour. I turned off my BlackBerry—this was tough, but I finally learned after my heart attack that no one is indispensible. My daughter scored two goals. I saw her kick in both; she waved at me after the second. My heart leaped! We're having family dinner tonight—can't wait.	Community, Oxytocin Vagal tone—high Fear center—low Rational frontal lobes—low
7:00 P.M.	Had a raucous chaotic dinner; it was horrible. I made the pasta. We ordered in finally. Great fun. Excused myself at 8:30 to work on e-mails.	Community, Oxytocin Vagal tone—high Fear center—low Rational frontal lobes—low
9:30 P.M.	Turned in early with my wife. She has a rule—no television in bedrooms. And now I don't bring my BlackBerry or laptop in, either.	Community, Oxytocin Vagal tone—high Fear center—low Rational frontal lobes—low
10:30 P.M.	We talk and fall asleep talking. Think it must be around 10:30. It's been a scary yet amazing day. I feel very blessed.	Natural, sufficient sleep Vagal tone—high Fear center—low Rational frontal lobes—low

Even if it seems implausible, this breaking up of Clifford's hectic schedule with calm-promoting behavior would surely have helped stave off his first heart attack and prevent future ones. It would keep him off antidepressants and improve his and his family's happiness. My "new improved" schedule would also help him become a more creative and nimble problem solver.

Most of us spend a significant part of our lives at work, and where we work and how we work have implications for our level of calm. Clifford is the CEO of a large company, but I believe that even if you are a small part of a large organization, it is possible to build calming moments into your work life. Let me share with you a bit about how I have deliberately brought core-brain calm into my workplace.

The Workplace

I have a general neurology practice, with a subspecialty in dementia, primarily Alzheimer's. About 60 percent of my practice involves memory loss. You can imagine how stressful it is for a patient and for a family coming in to have their memory evaluated and possibly hear a diagnosis they dread. So my goal has been to make my office as warm and comforting as possible, both for my patients and their families and for all of us who work here at the office. This is very important, because if the staff is stressed out or anxious, that surely communicates itself to patients and makes them even more nervous. In hiring staff, I look especially for a cheerful, friendly manner.

When I set up the practice, I chose to make it very much like a home. I didn't buy the furniture or fixtures at a medical-supply store, except for my examining table. The chairs are chairs that you would find in any living room, the art is what I would choose to hang at home. The lighting is muted. The front desk is not barricaded behind a counter or sequestered in a cubicle; it's out in the open, and patients can talk comfortably to the receptionist.

When you arrive at the office, you are not handed a clipboard with forms to fill out. Instead, you are greeted warmly and offered coffee or tea. I'm a big believer in chocolate as comfort food, so there's always some of that in one form or other on a plate. We have a little sunroom in the back for patients and families to relax in. Two large dogs pad around the office, unless of course someone is afraid of dogs, in which case they stay put behind my desk. The dogs do wonders for the oxytocin levels of all of us who work there, as well as for the patients. Dogs are excellent calm generators. One of our dogs recently passed a therapy training program, allowing her to offer her special brand of medicine to nursing homes and hospitals.

I remember once having to tell a patient that he had an inoperable brain tumor. It was a rainy day, and the next time he came to see me, he said that what he remembered most about that visit was how Max, one of the dogs, lay down on top of his wet feet and warmed them as I gave him the terrible news. Both Max and my patient are long gone, and I hope they are happy somewhere in a calm heaven.

Many of my staff and I sit on big exercise balls instead of chairs. The balls allow for a certain amount of playfulness, a certain amount of mobility, and they help with posture and back pain. Once or twice a week, we do Pilates exercises at the end of the day. This helps us laugh our tensions away and to bond as a group. Staff meetings, although rare (I'm still working on improving my management skills), are often held outside the office so that the threat level is down and creative ideas can flow.

In one of the staff meetings, we came up with the idea of an interoffice messaging system; now when one of us is in a stressful situation, we can send out a silent scream to a sympathetic colleague. All these small details, taken together, contribute to feelings of good humor and calm.

When I am with patients, even if it is for a few minutes, I find it calming for me to give them a hug or just a small touch. I believe this helps to calm them, too. So much of illness today has a component of anxiety, and these simple gestures, when genuine, go a long way toward helping someone feel better, even if the person is suffering from a brain tumor. A colleague of mine, a well-known and respected internist, claims that nearly 80 percent of his patients—the "worried well"—have enough anxiety to affect their health.

My practice is relatively small, with six staff members in all, but large organizations can also create a climate of calm to improve productivity and reduce angst. Consider Google's now famous people-friendly workplace. While many things Californian are "new think," Google has also achieved a lot of "core think."

Google is the fourth best company in the United States at which to work, according to both *Fortune* and *Forbes;* not surprisingly, close to three quarters of a million men and women apply for jobs there each year. Here's the way a recent *Fortune* magazine article describes the work environment.

> At Google you can do your laundry; drop off your dry cleaning; get an oil change, then have your car washed; work out in the gym; attend subsidized exercise classes; get a massage; study Mandarin, Japanese, Spanish, and French; and ask a personal concierge to arrange dinner reservations. Naturally you can get haircuts onsite.
>
> Want to buy a hybrid car? The company will give you $5,000 toward that environmentally friendly end. Care to refer a friend to work at Google? Google would like that, too, and it'll give you a $2,000 reward. Just have a new baby? Congratulations! Your employer will reimburse you for up to $500 in takeout food to ease your first four weeks at home. Looking to make new friends? Attend a weekly TGIF party, where there's usually a band playing. Five onsite doctors are available to

> give you a checkup, free of charge. There are lactation pumps for nursing mothers so they don't have to tote these back-and-forth from home.

The author of the article, Adam Lashinsky, chronicled Google's unique efforts to harness their employees' creativity. This included his description of sitting on a heated toilet seat on "pajama day" in engineering building number 40. A document hung in the toilet stall concerned with engineering conundrums such as "lode coverage." We don't know if any patents emerged from this particular pajama day, but . . .

Sitting on the heated toilet seat in your PJs means you have high vagal tone, low activity in the amygdala, and a lulled dorsolateral prefrontal cortex, a perfect setting for creativity.

Lashinsky made this broad observation of employees:

> Googlers tend to be happy-go-lucky on the outside but type A at their core. Ask one what he or she is doing, and it's never "selling ads" or "writing code." No, they're on a quest "to organize the world's information and make it universally accessible and useful."

This sort of gung ho attitude can get annoying no doubt. But there is something worth emphasizing here. Tony

Delamothe, an editor at the *British Medical Journal,* made the following observation in a 2005 editorial. "Work is central to well-being, and certain features correlate highly with happiness. These include autonomy over how, where, and at what pace work is done; trust between employer and employee; procedural fairness; and participation in decision-making." It seems that Googlers, thanks to cofounders Sergey Brin and Larry Page, enjoy all this and free sushi besides.

That *Fortune* article posed an intriguing question, at least to many investors: *"Is Google a great place to work because its stock is at $483, or is its stock at $483 because it's a great place to work?"* Knowing what you now know about the neuroscience of calm and productivity, you be the judge.

The Sound of One Friend Laughing

If we find some of the happiest workers on the Google campus in California, in which countries might we find the happiest citizens? Another business magazine publisher, Forbes, has investigated this question. In a survey of forty industrialized nations, Denmark not only came out on top as the happiest of the countries, but it also scored as the most productive. Norway, the richest nation, came in ninth. The United States, in case you're wondering, did not make the top ten. In the survey, happiness was determined by the way a thousand randomly selected people answered questions like "Did you enjoy something you did yesterday?" One important factor was work-family balance. In the Scandinavian countries, which were rated as most happy, people worked an average of thirty-seven hours a week. Obviously, working less does not explain happiness; there has to be a fulfilling home life as well.

Although the research found that financial security was essential to a basic level of happiness, and unemployment was highly correlated with unhappiness, it discerned little change in happiness in developed countries even as their economies improved. In fact, countries as economically dissimilar as Mexico and the United States report similar levels of happiness despite vast disparities in income. Economists talk about social capital, which is econospeak for social ties with other human beings, and this is more important to the brain—for calm and for an overall sense of well-being—than material capital.

It may be relative income rather than absolute income that makes the difference. This is consistent with evolutionary theory, because it's competition with members of the same community, in this case other Misters Joneses competing for the soon to be Mistress Jones (or other Senors Rodrigos competing for the soon to be Senora Rodrigo) that determines survival of the genes of Mister Jones and Senor Rodrigo.

There are cultural aspects to calm as well. Siestas are more common in some societies, for example. Western philosophies tend to approach happiness and calm in a more cerebral way than their Eastern counterparts. But from a neuroscientific perspective, Eastern practices, such as yoga and meditation, are particularly successful at creating calm, because they recruit core-brain circuits. This means they activate the parasympathetic and vagus nerve systems, promoting relaxation and toning down the energizing sympathetic nervous system. This soothing effect may be why so many people in the West have become so enamored of these practices.

A fascinating variation of yoga called laughter yoga, based on the breathing technique *kapalbhati*, has become increasingly popular in recent years. "Laughing clubs" have sprung up all over the world. "This is not a laughing matter," says Madan Kataria, one of the founders of these clubs, on National Public Radio. "It is a very serious thing that we've forgotten to laugh." Although it may sound silly for a group of people to get together and laugh themselves silly, there is a powerful core-brain rationale for this exercise. It activates our sense of community and our mirror neurons and insula, leaving us invigorated. If you doubt me, watch a bunch of toddlers dissolving into giggles over nothing in particular.

A vast body of literature speaks to the uplifting and healing powers of laughter. A friend of mine who attended a yoga laughing workshop at a psychology conference reports that after her initial self-consciousness and doubt, she got into it and left the room feeling "oddly energized." To her surprise, that feeling remained with her the rest of the day.

Today's Steps to a Calm Brain

I began this book talking about my grandfather, my beloved Thatha, who showed me by quiet example why it is so important to integrate calm into our lives. For him, vitality and peacefulness coexisted harmoniously. His was a life of balance, the result of a calm brain.

I hope you now have the essential insights of neurology you need to create your own calm balance and unlock your natural relaxation system—never more necessary than in our era of digital overload and urban disconnection. Understanding the neural underpinnings of calm—the delicate dance of the emotional core brain and the rational frontal lobes—is the first step. With a firm grasp of the complexities of calm, we can all begin to embrace it in the deliberate and confident way that my grandfather did.

Increasing vagal tone and amping up the parasympathetic relaxing system mean connecting with our communities and enjoying a fulfilling work and home life. While this may sound abstract or out of reach, it is in fact concrete, and steps toward attaining it are there for the taking—right now, today.

I'd like to end our journey with a list of practical suggestions. Laughter is an easy way to enhance vagal tone; making an intentional effort to laugh every day has been scientifically shown to promote relaxation, increase happiness, and improve health. Here are some other practices that will deliver calm from the bottom of your brain all the way to the top.

- Form long-lasting, close ties with other human beings. (And laugh with them.)
- Find a life partner who offers both companionship and romance.
- Have sex as often as you can; it's good for your core brain.
- Forgive and forget. As difficult as this may seem, it is worth the effort. Forgiveness makes us less bitter and more friendly throughout our lives—especially in our later years. Forgiving helps restore a sense of community.
- Pay attention to your biorhythms: Eat when you are hungry, sleep when you are tired, wake up when you are rested.
- Never, ever, compromise on sleep. This is the easiest, best thing you can do for your core-brain calm this week. (Make a weekend of it—turn in Friday, turn off electronic gadgets, and retire to your bed. Welcome to your sleepcation!)
- Stop using drugs to stay awake and to get to sleep and alarms to wake up. Try to figure out

your natural sleep cycle on a weekend (when you aren't on a sleepcation) and use that to determine your bedtimes and your wake-up times.

- Cut down on multitasking. While doing many things at once is fast becoming the norm, relegate it to certain periods of time.
- Schedule some downtime in the middle of hectic days.
- Stop watching television. Hey, even get rid of it.
- Do one slow thing a day. Walk to work, stop and pet a dog on the street, smile at a stranger.
- Practice real communication. Virtual communication (via e-mail or texting) is not the same as face-to-face communication. Your core brain makes the distinction. With virtual communication, even with telephones, significant subtle signals are lost, including core-brain triggers like smell, and this makes for a core-brain disconnect. Spend time socializing without your cell phone next to you.
- Hug someone you care about as much and as often as you can; communicate with touch.

- Pets are great sources of calm. Make them a habit you can't live without.
- And, finally, like my grandfather, welcome each morning calmly, slowly pouring your mind and body into the day.

The tapestry of calm you weave will be your own.

References and Suggested Additional Reading

Occasional articles are listed more than once for ease of reference.

Chapter 1

Belouin SJ, Reuter N, Borders-Hemphill V, Mehta H. Prescribing trends for opioids, benzodiazepines, amphetamines, and barbiturates from 1998–2007. nac.samhsa.gov.

Dawkins R. *Selfish Gene*, new ed. Oxford: Oxford University Press, 1989.

Fuster JM. Frontal lobe and cognitive development. *J Neurocytology* 2002; 3:373–385.

Gazzaniga MS. *The Ethical Brain: The Science of Our Moral Dilemmas*. Harper Perennial, 2006.

Hale TA. *Griots and Griottes: Masters of Words and Music*. Bloomington, Indiana: Indiana University Press, 1998.

Hill RA, Dunbar RIM. Social network size in humans. *Hum Nat.* 2003; 14:53–72.

Lambert KG. Rising rates of depression in today's society: consideration of the roles of effort-based rewards and enhanced resilience in day-to-day functioning. *Neurosci Biobehavioral Rev.* 2006; 30:497–510.

Manchikanti L, Fellow B, Ailinani H, Pampati V. Therapeutic use, abuse, and nonmedical use of opioids: a ten-year perspective. *Pain Physician* 2010; 13:401–435.

Martin EI, Ressler KJ, Binder E, Nemeroff CB. The neurobiology of anxiety disorders: brain imaging, genetics, and psychoneuroendocrinology. *Psychiatr Clin N Amer.* 2009; 32:549–575.

Northcutt RG, Kaas JH. The emergence and evolution of mammalian neocortex. *Trends Neurosci* 1995; 18:373–379.

Seymour B, Singer T, Dolan R. The neurobiology of punishment. *Nat Rev Neurosci.* 2007; 8:300–311.

Twenge, J, Gentile B, DeWall CN, et al. Birth cohort increases in psychopathology among young Americans, 1938–2007: a cross-temporal meta-analysis of the MMPI. *Clin Psychology Rev.* 2010; 30: 145–154.

Chapter 2

Blechert J, Michael T, Grossman P, et al. Autonomic and respiratory characteristics of posttraumatic stress disorder and panic disorder. *Psychosom Med.* 2007; 69:935–943.

Bowsher D, Geoffrey Woods C, Nicholas AK, et al. Absence of pain with hyperhidrosis: a new syndrome where vascular afferents may mediate cutaneous sensation. *Pain* 2009; 147: 287–298.

Bracha HS. Human brain evolution and the "Neuroevolutionary time-depth principle": implications for the reclassification of fear-circuitry-related traits in DSM-V and for studying resilience to warzone-related posttraumatic stress disorder. *Prog NeuroPsychopharm Biol Psychiatry* 2006; 30: 827–853.

Breiter HC, Etcoff NL, Whalen PJ, et al. Response and habituation of the human amygdala during visual processing of facial expression. *Neuron* 1996; 17:875–887.

Darwin C. *The Expression of Emotions in Man and Animals*. New York: Appleton, 1872.

Diamond J. "Vengeance Is Ours." *The New Yorker*, April 21, 2008; 74–87.

Gray H. *Anatomy of the Human Body*. Philadelphia: Lea & Febiger, 1918.

Nielsen JM, Sedgwick RP. Instincts and emotions in an anencephalic monster. *J Nervous Mental Dis.* 1949; 110:387–394.

Paton WDM. The Hexamethonium Man. *Pharm Rev.* 1954; 6:59.

Porges SW. The polyvagal perspective. *Biol Psychol.* 2007; 74:116–143.

Porges SW. The polyvagal theory: new insights into adaptive reactions of the autonomic nervous system. *Clev Clin J Med.* 2009; 76:S86–S90.

Saper CB. Pattern generators for conveying emotion: the central autonomic nervous system: conscious visceral perception and autonomic pattern generation. *Ann Rev Neurosci.* 2002; 25: 433–469.

Semendeferi K, Damasio H. The brain and its main anatomical subdivisions in living hominoids using magnetic resonance imaging. *J Hum Evol.* 2000; 38:317–332.

Swanson LW. What is the brain? *Trends Neurosci.* 2000; 23: 519–527.

Zagon A. Does the vagus nerve mediate the sixth sense? *Trends Neurosci.* 2001; 24:671–673.

Chapter 3

Bernardi L, Sleight P, Bandinelli G, et al. Effect of rosary prayer and yoga mantras on autonomic cardiovascular rhythms: a comparative study. *Br Med J* 2001; 323:22–29.

Brown RP, Gerbarg PL. Sudarshan Kriya yogic breathing in the treatment of stress, anxiety, and depression: Part I—Neurophysiologic model. *J Altern Complement Med.* 2005; 11:189–201.

Craig ADB. Interoception: the sense of the physiological condition of the body. *Curr Opin Neurobiol.* 2003; 13:500–505.

Das A. "Scared to Death." *Boston Globe*, August 6, 2006.

Friedman BH. An autonomic flexibility–neurovisceral integration model of anxiety and cardiac vagal tone. *Biological Psychology* 2007; 74:185–199.

George MS, Sackeim HA, Rush AJ, et al. Vagus nerve stimulation: a new tool for brain research and therapy. *Biol Psychiatry* 2000; 47:287–295.

Ghizoni DM, Joao LM, Neto LM, et al. The effects of metabolic stress and vagotomy on emotional learning in an animal model of anxiety. *Neurobiol Learning Memory* 2006; 86:107–116.

Goehler LE, Lyte M, Gaykema PA. Infection-induced viscerosensory signals from the gut enhance anxiety: implications for psychoneuroimmunology. *Brain Behav Immun.* 2007; 21: 721–726.

Gray MA, Rylander K, Harrison NA, et al. Following one's heart: cardiac rhythms gate central initiation of sympathetic reflexes. *J Neurosci.* 2009; 29:1817–1825.

Mezzacappa ES, Kelsey RM, Katkin ES, et al. Vagal rebound and recovery from psychological stress. *Psychosom Med.* 2001; 63:650–657.

Miu AC, Heilman RM, Miclea M. Reduced heart rate variability and vagal tone in anxiety: trait versus state, and the effects of autogenic training. *Autonom Neurosci: Basic Clin.* 2009; 145: 99–103.

Pegna AJ, Khateb A, Lazeyras F, Seghier ML. Discriminating emotional faces without primary visual cortices involves the right amygdala. *Nat Neurosci.* 2005; 8:24–25.

Rizzolatti G, Craighero L. The mirror neuron system. *Ann Rev Neurosci.* 2004; 27:169–192.

Rosenkranz MA, Busse WW, Johnstone T., et al. Neural circuitry underlying the interaction between emotion and asthma symptom exacerbation. *Proc Nat Acad Sci.* 2005; 102:13319–13324.

Rottenberg J, Clift A, Bolden S, Salomon K. RSA fluctuation in major depressive disorder. *Psychophysiol.* 2007; 44:450–458.

Samuels MA. Contemporary reviews in cardiovascular medicine: the brain–heart connection. *Circulation* 2007; 116:77-84.

Saper CB, Iversen S, Frackowiak R. "Integration of Sensory and Motor Function: The Association Areas of the Cerebral Cortex and the Cognitive Capabilities of the Brain." In: Kandel E, Schwartz J, Jessell T., editors. *Principles of Neural Science*, 4th ed. McGraw-Hill Medical, 2000: 349–380.

Tsuchiya N, Moradi F, Felsen C, et al. Intact rapid detection of fearful faces in the absence of the amygdala. *Nat Neurosci.* 2009; 12:1224–25.

Yeragani VK, Tancer M, Seema KP, et al. Increased pulse-wave velocity in patients with anxiety: implications for autonomic dysfunction. *J Psychosom Research* 2006; 61:25– 31.

Chapter 4

Arnsten AF. Stress signalling pathways that impair prefrontal cortex structure and function. *Nat Rev Neurosci.* 2009; 10:410–422.

Blásquez JCC, Font GR, Ortís LC. Heart-rate variability and precompetitive anxiety in swimmers. *Psicothema* 2009; 21: 531–536.

Csikszentmihalyi M. *Flow: The Psychology of Optimal Experience.* Harper Perennial, 1991: 320.

Faucher L, Blanchette I. *Fearing New Dangers: Phobias and the Cognitive Complexity of Human Emotions.* New York: Oxford University Press, 2011.

Feder A, Nestler EJ, Charney DS. Psychobiology and molecular genetics of resilience. *Nat Rev Neurosci.* 2009; 10: 446–457.

Fishbane MD. Wired to connect: neuroscience, relationships, and therapy. *Family Process* 2007; 46:395–412.

Glover H. Emotional numbing: a possible endorphin-mediated phenomenon associated with post-traumatic stress disorder and other allied psychopathological states. *J Traumatic Stress* 1992; 5:643–675.

Hoffmann AA, Harshman LG. Desiccation and starvation resistance in Drosophila: patterns of variation at the species, population and intrapopulation levels. *Heredity* 1999; 83:637–643.

Joëls M, Baram TZ. The neuro-symphony of stress. *Nat Rev Neurosci.* 2009; 10:459–466.

Kaati G, Bygren LO, Edvinsson, S. Cardiovascular and diabetes mortality determined by nutrition during parents' and grandparents' slow growth period. *Eur J Hum Genet* 2002; 10:682–688.

Leonard WR, Snodgrass JJ, Robertson ML. Effects of brain evolution on human nutrition and metabolism. *Ann Rev Nutrition* 2007; 27:311–327.

Livingstone D. *Missionary Travels and Researches in South Africa, Including a Sketch of Sixteen Years' Residence in the Interior of Africa, and a Journey from the Cape of Good Hope to Loanda on the West Coast, Thence Across the Continent, Down the River Zambesi, to the Eastern Ocean.* London: Ward, Lock & Co., 1857.

Marks IM, Nesse RM. Fear and fitness: an evolutionary analysis of anxiety disorders. Special issue: Mental disorders in an evolutionary context. *Ethol Sociobiol* 1999; 15:247–261.

Nesse RM, Young EA. "Evolutionary Origins and Functions of the Stress Response." In: Fink G, editor. *Encyclopedia of Stress.* Vol 2, pp. 79–84. San Diego: Academic Press, 2000.

Nesse RM, Williams GC. *Why We Get Sick: The New Science of Darwinian Medicine.* Vintage, 1996.

Pembrey M, Bygren LO, Kaati GP, et al. Sex-specific, sperm-mediated transgenerational responses in humans. *Eur J Hum Genet* 2005; 14:159–166.

Pennington BF. *The Development of Psychopathology: Nature and Nurture.* New York: Guilford Press, 2005; 380.

Sheline YI, Barch DM, Price JL, et al. The default mode network and self-referential processes in depression. *Proc Nat Acad Sciences* 2009; 106:1942–1947.

Susser ES, Lin SP. Schizophrenia after prenatal exposure to the Dutch Hunger Winter of 1944–1945. *Arch Gen Psychiatry* 1992; 49:983–988.

The Churchill Centre and Museum at the Churchill War Rooms, London. www.winstonchurchill.org.

Weaver ICG, Ceryoni N, Champagne FA. Epigenetic programming by maternal behavior. *Nat Neurosci.* 2004; 7:847–854.

Whitelaw E. Epigenetics: Sins of the fathers, and their fathers. *Eur J Human Genetics* 2006; 14:131–132.

Chapter 5

Bargh JA, Chen M, Burrows L. Automaticity of social behavior: direct effects of trait construct and stereotype activation on action. *J Personality Social Psychology* 1996; 71:230–244.

Bechara A. Decision-making, impulse control and loss of willpower to resist drugs: a neurocognitive perspective. *Nat Neurosci.* 2005; 8:1458–1463.

Bota M, Dong HW, Swanson LW. From gene networks to brain networks. *Nat Neurosci.* 2003; 6:795–799.

Brefczynski-Lewis JA, Lutz A, Schaefer HS, et al. Neural correlates of attentional expertise in long-term meditation practitioners. *Proc Nat Acad Sciences* 2007; 104:11483–11488.

Brickman P, Coates D, Janoff-Bulman R. Lottery winners and accident victims: is happiness relative? *J Personality Social Psychology* 1978; 36:917–927.

Buschman TJ, Miller EK. Top-down versus bottom-up control of attention in the prefrontal and parietal cortices. *Science* 2007; 315:1860–1862.

Carroll SB. Genetics and the making of Homo sapiens. *Nature* 2003; 422:849–857.

Dapretto M, Davies MS, Pfeifer JH, et al. Understanding emotions in others: mirror neuron dysfunction in children with autism. *Nat Neurosci.* 2005; 9:28–30.

Davidson RJ, Maxwell JS, Shackman AJ. The privileged status of emotion in the brain. *Proc Nat Acad Sciences* 2004; 101:11915–11916.

Dolan RJ. Neuroimaging of cognition: past, present, and future. *Neuron* 2008; 60:496–502.

Eisenberger NI, Lieberman MD, Williams KD. Does rejection hurt? An FMRI study of social exclusion. *Science* 2003; 302:290–292.

Fischer AH, Rotteveel M, Evers C, Manstead ASR. Emotional assimilation: how we are influenced by others' emotions. *Cahier de Psychologie Cognitive* 2004; 22:223–245.

Flinn MV, Geary DC, Ward CV. Ecological dominance, social competition, and coalitionary arms races: why humans evolved extraordinary intelligence. *Evol Hum Behav.* 2005; 26:10–46.

Gilbert, D. *Stumbling on Happiness*. Knopf, 2006. 304 pg.

Gladwell M. "On Spaghetti Sauce." TED Talks; www.ted.com.

Iacoboni M, Dapretto M. The mirror neuron system and the consequences of its dysfunction. *Nat Rev Neurosci.* 2006; 7:942–951.

Kneller A. Transitivity, the orbitofrontal cortex, and neuroeconomics. *Harvard Gazette*, December 10, 2007.

Knight RT. Neuroscience—Neural networks debunk phrenology. *Science* 2007; 316:1578–1579.

Koechlin E, Charron S. Divided representations of concurrent goals in the human frontal lobes. *Science* 2010; 328:360–363.

Kornhuber HH, Deecke L. Hirnpotentialänderungen bei Willkürbewegungen und passiven Bewegungen des Menschen: Bereitschaftspotential und reafferente Potentiale. *Pflügers Arch* 1965; 284:1–17.

Krubitzer L. In search of a unifying theory of complex brain evolution. *Ann NY Acad Sci.* 2009; 1156:44–67.

Leary MR, Tate EB, Adams CE, et al. Personality processes and individual differences: self-compassion and reactions to unpleasant self-relevant events: the implications of treating oneself kindly. *J Personality Soc Psychol* 2007; 92:887–904.

Longe O, Maratos FA, Gilbert P, et al. Having a word with yourself: neural correlates of self-criticism and self-reassurance. *NeuroImage* 2010; 49:1849–1856.

McEwen, B. The brain is the central organ of stress and adaptation. *Neuroimage* 2009; 47:911–913.

Meltzoff AN. "Imitation and Other Minds: The 'Like Me' Hypothesis." In: Hurley S, Chater N, editors. *Perspectives on Imitation: From*

Cognitive Neuroscience to Social Science. Vol. 2, pp. 55–77. Cambridge: MIT Press, 2005.

Miller EK, Cohen JD. An integrative theory of prefrontal cortex function. *Ann Rev Neurosci.* 2001; 24:167–202.

Muzur A, Pace-Schott EF, Hobson AJ. The prefrontal cortex in sleep. *Trends Cogn Sci.* 2002; 6:475–481.

Nesse RM. Natural selection and the regulation of defenses: a signal detection analysis of the smoke detector principle. *Evol Hum Behav* 2005; 26:88–105.

Rizzolatti G, Craighero L. The mirror neuron system. *Ann Rev Neurosci*. 2004; 27:169–192.

Robinson GE, Fernald RD, Clayton DF. Genes and social behavior. *Science* 2008; 322:896–900.

Schwartz B. *The Paradox of Choice: Why More Is Less.* Ecco, 2003. 288 pg.

Swanson LW. Anatomy of the soul as reflected in the cerebral hemispheres: neural circuits underlying voluntary control of basic motivated behaviors. *J Comparative Neurol* 2005; 493:122–131.

Welberg L. Learning to reflect. *Nat Rev Neurosci.* 2007; 8:737.

Chapter 6

Abbott A. "Blind Man Walking." *Nature News*, December 22, 2008.

Bechara A, Damasio AR. The somatic marker hypothesis: a neural theory of economic decision. *Games Econ Behav.* 2005; 52: 336–372.

Bechara A. Editorial, The neurology of social cognition. *Brain* 2002; 125:1673–1675.

Craig AD. Perspectives: how do you feel? Interoception: the sense of the physiological condition of the body. *Neurosci Nat Rev* 2002; 3:655–665.

Craig ADB. Interoception: the sense of the physiological condition of the body. *Curr Opin Neurobiol* 2003; 13:500–505.

Damasio AR. Emotion in the perspective of an integrated nervous system. *Brain Research Rev* 1998; 26:83–86.

Farb NAS, Anderson AK, Mayberg H, et al. Minding one's emotions: mindfulness training alters the neural expression of sadness. *Emotion* 2010; 10:25–33.

Harrison NA, Brydon L, Walker C, et al. Neural origins of human sickness in interoceptive; responses to inflammation. *Biol Psychiatry* 2009; 66:415–422.

Harrison NA, Critchley HD. Affective neuroscience and psychiatry. *Br J Psychiatry* 2007; 191:192–194.

Kluver H, Bucy PC. Preliminary analysis of functions of the temporal lobes in monkeys. *Arch Neurol Psych.* 1939; 42:979–1000.

Lutz A, Brefczynski-Lewis J, Johnstone T, Davidson RJ. Regulation of the neural circuitry of emotion by compassion meditation: effects of meditative expertise. *PLoS ONE* 2008; 3: e1897. doi:10.1371/journal.pone.0001897.

Mayberg HS, Lozano A, Voon V, et al. Deep brain stimulation for treatment resistant depression. *Neuron* 2005; 45:651–660.

Morris JS, Ohman A, Dolan RJ. A subcortical pathway to the right amygdala mediating "unseen" fear. *Proc Nat Acad Sciences* 1999; 96:1680–1685.

Olds J, Milner P. Positive reinforcement produced by electrical stimulation of septal area and other regions of rat brain. *J Comp Physiol Psychol.* 1954; 47:419–427.

Pegna AJ, Khateb A, Lazeyras F, Seghier ML. Discriminating emotional faces without primary visual cortices involves the right amygdala. *Nat Neurosci.* 2005; 8:24–25.

Ressler KJ, Mayberg HS. Targeting abnormal neural circuits in mood and anxiety disorders: from the laboratory to the clinic. *Nat Neurosci.* 2007; 10:1116–1124.

Rogers K, Schlitz D. "The Gambler." United Artists, 1978.

Roozendaal B, McEwen BS, Chattarji S. Stress, memory and the amygdala. *Nat Rev Neuroscience* 2009; 10:423–433.

Scheuerecker J, Frodl T, Koutsouleris N, et al. Cerebral differences in explicit and implicit emotional processing: an fMRI study. *Neuropsychobiology* 2007; 56:32–39.

Stefurak T, Mikulis D, Mayberg H, et al. Deep brain stimulation for Parkinson's disease dissociates mood and motor circuits: a functional MRI case study. *Mov Disord* 2003; 18:1508–1516.

Terzian H, Ore GD. Syndrome of Kluver and Bucy: reproduced in man by bilateral removal of the temporal lobes. *Neurology* 1955; 5:373–380.

Verdejo-Garcia A, Perez-Garcia M, Bechara A. Emotion, decision-making and substance dependence: a somatic-marker model of addiction. *Curr Neuropharm* 2006; 4:17–31.

Chapter 7

Aserinsky E, Kleitman N. Regularly occurring periods of eye motility, and concomitant phenomena, during sleep. *Science* 1953; 118:273–274.

Bergmann BM, Everson CA, Kushida CA, et al. Sleep deprivation in the rat: V. Energy use and mediation. *Sleep* 1989; 12:31–41.

Bergmann BM, Kushida CM, Everson EA, et al. Sleep deprivation in the rat: II. Methodology. *Sleep* 1989; 12:5–12.

Bergmann BM, Kushida CM, Everson EA, et al. Sleep deprivation in the rat: III. Total sleep deprivation. *Sleep* 1989; 12:13–21.

Deregnaucourt S, Mitra PP, Feher O, et al. How sleep affects the developmental learning of bird song. *Nature* 2005; 433: 710–716.

Ericsson KA, Krampe RT, Tesch-Romer C. The role of deliberate practice in the acquisition of expert performance. *Psychol Rev.* 1993; 100:363–406.

Everson CA, Bergmann BM, Rechtschaffen A. Sleep deprivation in the rat: VI. Skin changes. *Sleep* 1989;12:42–46.

Everson CA, Toth LA. Systemic bacterial invasion induced by sleep deprivation. *Am J Physiol Regul Integr Comp Physiol* 2000; 278:R905–916.

Gellene D. "Sleeping Pill Use Grows as Economy Keeps People Up at Night." *LA Times*, March 30, 2009.

He Y, Jones CR, Fujiki N, et al. The transcriptional repressor DEC2 regulates sleep length in mammals. *Science* 2009; 325: 866–870.

Horvath TL, Gao XB. Input organization and plasticity of hypocretin neurons: possible clues to obesity's association with insomnia. *Cell Metabolism* 2005; 1:279–286.

Kollar EJ, Pasnau RO, Rubin RT, et al. Psychological, psychophysiological, and biochemical correlates of prolonged sleep deprivation. *Amer J Psychiatry* 1969; 126:488–497.

Luby ED, Frohman CF, Griswell JL, et al. Sleep deprivation: effects on behavior, thinking, motor performance, and biological energy transfer systems. *Psychosom Med.* 1960; 22:183–192.

Lutz A, Greischar LL, Rawlings NB, et al. Long-term meditators self-induce high-amplitude gamma synchrony during mental practice. *Proc Nat Acad Sci.* 2004; 101:16369–16373.

Nicholson AN, Stone BM, Clarke CH. Effect of the 1,5-benzodiazepines, clobazam and triflubazam, on sleep in man. *Br J Clin Pharmacol.* 1977; 4:567–572.

Pasnau RO, Naitoh P, Stier S, Kollar EJ. The psychological effects of 205 hours of sleep deprivation. *Arch Gen Psychiatry* 1968; 18: 496–505.

Rechtschaffen A, Siegel J. "Sleep and Dreaming." In: Kandel E, Schwartz J, Jessell T., editors. *Principles of Neural Science*, 4th edition. McGraw-Hill Medical, 2000: 936–959.

Roth T, Zorick F, Sicklesteel J, Stepanski E. Effect of benzodiazepines on sleep and wakefulness. *Br J Clin Pharmacol* 1981; 11 Suppl 1:31S–35S.

Siegel JM. Do all animals sleep? *Trends Neurosci.* 2008; 31:208–213.

Siegel JM. Sleep viewed as a state of adaptive inactivity. *Nat Rev Neurosci.* 2009; 10:747–753.

Smaldone A, Honig JC, Byrne MW. Sleepless in America: inadequate sleep and relationships to health and well-being of our nation's children. *Pediatrics* 2007; 119: S29–S37.

Stickgold R, Ellenbogen JM. "Sleep on It: How Snoozing Makes You Smarter." (During slumber, our brain engages in data analysis, from strengthening memories to solving problems.) *Scientific American Mind*, August 7, 2008.

Taylor DJ, Lichstein KL, Durrence HH, et al. Epidemiology of insomnia, depression, and anxiety. *Sleep* 2005; 28(11):1457–1464.

Chapter 8

Baumgartner T, Heinrichs M, Vonlanthen A, et al. Oxytocin shapes the neural circuitry of trust and trust adaptation in humans. *Neuron* 2008; 58:639–650.

Bielecka-Dabrowa A, Mikhailidis DM, Hannam S, et al. Takotsubo cardiomyopathy—the current state of knowledge. *Int J Cardiol.* 2010; 142:120–125.

Buss DM, Larsen RJ, Westen D, et al. Sex differences in jealousy: evolution, physiology and psychology. *Psychol Sci.* 1992; 3: 251–255.

Cahill L. "His Brain, Her Brain." *Scientific American*, May 2005: 292(5): 40-47.

Carr PL, Ash AS, Friedman RH, et al. Faculty perceptions of gender discrimination and sexual harassment in academic medicine. *Ann Intern Med.* 2000; 132:889–896.

Chaviaras S, Mak P, Ralph D, et al. Assessing the antidepressant-like effects of carbetocin, an oxytocin agonist, using a modification of the forced swimming test. *Psychopharm* (Berl). 2010; 210: 35–43.

Decety J, Jackson PL, Sommerville JA, et al. The neural bases of cooperation and communication: an fMRI investigation. *Neuroimage* 2004; 23:744–751.

Devi G, Hahn K, Massimi S, Zhivotovskaya E. Prevalence of memory loss complaints and other symptoms associated with the menopause transition: a community survey. *Gender Med.* 2005; 2:255–264.

Ditzen B, Schaer M, Gabriel B, et al. Intranasal oxytocin increases positive communication and reduces cortisol levels during couple conflict. *Biol Psychiatry* 2009; 65:728–731.

Donaldson ZR, Young L. Oxytocin, vasopressin, and the neurogenetics of sociality. *Science* 2008; 322:900–904.

Frank E, Brogan D, Schiffman M. Prevalence and correlates of harassment among US women physicians. *Arch Intern Med.* 1998; 158:352–358.

Gardiner M, Tiggemann M. Gender differences in leadership style, job stress, mental health in male- and female-dominated industries. *J Occupational Organizational Psychol.* 1999; 72:301–315.

Gianni M, Dentali F, Grandi AM, et al. Apical ballooning syndrome or Takotsubo cardiomyopathy: a systematic review. *Eur Heart J.* 2006; 27:1523–1529.

Heim C, Young LJ, Newport DJ, et al. Lower CSF oxytocin concentrations in women with a history of childhood abuse. *Mol Psychiatry* 2009; 14:954–958.

Hurlemann R, Patin A, Onur OA, et al. Oxytocin enhances amygdala-dependent, socially reinforced learning and emotional empathy in humans. *J Neurosci.* 2010; 30:4999–5007.

Jood K, Redfors P, Rosengren A, et al. Self-perceived psychological stress and ischemic stroke: a case-control study. *BMC Medicine* 2009; 7:53.

Kilgore WDS, Oki M, Yurgelun-Todd DA. Sex-specific developmental changes in amygdala responses to affective faces. *Neuroreport* 2001; 12:427–433.

Kosfeld M, Heinrichs M, Zak PJ, et al. Oxytocin increases trust in humans. *Nature* 2005; 435:673–676.

Kuo TB, Lin T, Yang CC, et al. Effect of aging on gender differences in neural control of heart rate. *Am J Physiol.* 1999; 277: H2233–2239.

Lown B. Sudden cardiac death: biobehavioral perspective. *Circulation* 1987; 76:I186–196.

Piccinelli M, Wilkinson G. Gender differences in depression, a critical review. *Br J Psychiatry* 2000; 177:486–492.

Schernhammer ES, Colditz GA. Suicide rates among physicians: a quantitative and gender assessment (meta-analysis). *Am J Psychiatry* 2004; 161:2295–2302.

Stroud LR, Salovey P, Epel ES. Sex differences in stress responses: social rejection versus achievement stress. *Soc Biol Psychiatry* 2002; 52:318–327.

Taylor SE, Klein LC, Lewis BP, et al. Biobehavioral responses to stress in females: tend-and-befriend, not fight-or-flight. *Psychol Rev* 2000; 107:411–429.

Troppmann KM, Palis BE, Goodnight JE, et al. Women surgeons in the new millennium. *Arch Surg* 2009; 144:635–642.

Wang J, Korczykowski K, Rao H, et al. Gender difference in neural response to psychological stress. *Soc Cogn Affect Neurosci.* 2007; 2:227–239.

Yee JR, Frijling J, Saber M, et al. Oxytocin alters the behavioral, cardiovascular, and hormonal responses to a mild daily stressor. Neuroscience Meeting, San Diego, November 2010. Abstract 190.7.

Ziabreva I, Poeggel G, Schnabel R, Braun K. Separation-induced receptor changes in the hippocampus and amygdala of octodon degus: influence of maternal vocalizations. *J Neurosci.* 2003; 23:5329–5336.

Chapter 9

Cloud J. "Why Your DNA Isn't Your Destiny." *Time* Magazine, January 6, 2010.

Gaylor EE, Goodlin-Jones BL, Anders TF. Classification of young children's sleep problems: a pilot study. *J Am Acad Child Adolesc Psychiatry* 2001; 40:61–67.

Gazelle H, Druhen MJ. Anxious solitude and peer exclusion predict social helplessness, upset affect, and vagal regulation in response to behavioral rejection by a friend. *Dev Psychol.* 2009; 45:1077–1096.

Henderson HA, Marshall PJ, Fox NA, et al. Psychophysiological and behavioral evidence for varying forms and functions of nonsocial behavior in preschoolers. *Child Dev.* 2004; 75: 251–263.

Herschkowitz N, Kagan J, Zilles K. Neurobiological bases of behavioral development in the second year. *Neuropediatrics* 1999; 30:221–230.

Kagan J, Reznick JS, Clarke C, et al. Behavioral inhibition to the unfamiliar. *Child Dev.* 1984; 55:2212–2225.

Kagan J, Reznick JS, Snidman N. Biological bases of childhood shyness. *Science* 1988; 240:167–171.

Kagan J, Snidman N, McManis M, Woodward SA. Temperamental contributions to the affect family of anxiety. *Psychiat Clin N Amer.* 2001; 24: 677–688. Review.

Kagan J, Snidman N. Early childhood predictors of adult anxiety disorders. *Soc Biol Psychiatry* 1999; 46:1536–1541.

Kagan J. The role of parents in children's psychological development. *Pediatrics* 1999; 104:164–167.

Kennedy AE, Rubin KH, Hastings PD, Maisel B. Longitudinal relations between child vagal tone and parenting behavior: 2 to 4 years. *Dev Psychobiol* 2004; 45:10–21.

Lawrence DH. "The Rocking-Horse Winner." *Harper's Bazaar,* July 1926.

Melman S, Little SG, Akin-Little KA. Adolescent overscheduling: the relationship between levels of participation in scheduled activities and self-reported clinical symptomology. *The High School Journal* 2007; 90:18–30.

Perlman SB, Camras LA, Pelphrey KA. Physiology and functioning: parents' vagal tone, emotion, socialization, and children's emotion knowledge. *J Exp Child Psych.* 2008; 100:308–315.

Sadeh A, Mindell JA, Luedtke K, et al. Sleep and sleep ecology in the first 3 years: a web-based study. *J. Sleep Res.* 2009; 18:60–73.

Sarkar P, Bergman K, Fisk NM, et al. Ontogeny of foetal exposure to maternal cortisol using midtrimester amniotic fluid as a biomarker. *Clin Endocrin.* 2007; 66:636–640.

Stein MA, Mendelsohn J, Obermeyer WH, et al. Sleep and behavior problems is school-aged children. *Pediatrics* 2001; 107:1–9.

Supplee LH, Skuban EM, Shaw DS, et al. Emotion regulation strategies and later externalizing behavior among European

American and African American children. *Dev Psychopathol* 2009; 21:393–415.
Van den Bergh BRH, Marcoen A. High antenatal maternal anxiety is related to ADHD symptoms, externalizing problems and anxiety in 8- and 9-year-olds. *Child Dev* 2004; 75:1085–1097.

Chapter 10

Delamothe T. Happiness. *Br Med J* 2005; 331:1489–1490.
Lashinsky A. "Google Is No. 1: Search and Enjoy." *Fortune*, January 2007.
Levy F. "The World's Happiest Countries." *Forbes*, July 2010.

Index